II International Workshop "Thermal Methods for Enhanced Oil Recovery: Laboratory Testing, Simulation and Oilfields Applications" (ThEOR 2017)

IOP Conference Series: Earth and Environmental Science
Volume 155

Kazan, Russia
19 – 23 June 2017

ISBN: 978-1-5108-6338-5
ISSN: 1755-1307

Printed from e-media with permission by:

Curran Associates, Inc.
57 Morehouse Lane
Red Hook, NY 12571

Some format issues inherent in the e-media version may also appear in this print version.

Printed by Curran Associates, Inc. (2018)

For permission requests, please contact the Institute of Physics at the address below.

Institute of Physics
Dirac House, Temple Back
Bristol BS1 6BE UK

Phone: 44 1 17 929 7481
Fax: 44 1 17 920 0979

techtracking@iop.org

Additional copies of this publication are available from:

Curran Associates, Inc.
57 Morehouse Lane
Red Hook, NY 12571 USA
Phone: 845-758-0400
Fax: 845-758-2633
Email: curran@proceedings.com
Web: www.proceedings.com

Table of contents

Volume 155

Thermal Methods for Enhanced Oil Recovery: Laboratory Testing, Simulation and Oilfields Applications (ThEOR2017)

19–23 June 2017, Kazan, Russian Federation

Accepted papers received: 4 May 2018
Published online: 22 May 2018

Preface

Papers

IOP Conf. Series: Earth and Environmental Science **155** (2018) 011001 doi:10.1088/1755-1315/155/1/011001

About Theor 2017

It is our great pleasure to invite you to the II International Workshop "Thermal Methods for Enhanced Oil Recovery: Laboratory Testing, Simulation and Oilfields Applications", ThEOR2017, which will be held in Kazan, Russia. ThEOR2017 is aimed to bring together key researchers and professionals from the industry, R&D centers and academic institutes to share experiences, perspectives, insights and the latest developments in thermal methods for EOR from all over world. The workshop's program contains four days of training courses, plenary and key-note lectures, technical and poster sessions on cutting edge issues related to petroleum industry. At the same time, we tried to keep the balance between traditional and emerging fields, fundamentals and applications, modeling and experiment, academic interests and industry needs.

The first conference ThEOR2016 has gathered more than 150 registered participants from 45 different organizations all over the world. This fact gives us enormous satisfaction and pride, as this event attracted interest of the community and can be continued in 2017.

Sections

1.Steam Assisted Gravity Drainage (SAGD), including solvent-aided and reagent-aided processes
2.Catalytic Compositions for In-Situ Oil Upgrading (development, characterization and usage of different catalysts for pyrolysis, oxidation, hydrocracking, aquathermolysis of fossil fuels)
3.Modeling and Simulation of Thermal EOR, including development of new software
4.Production of Unconventional Resources (tight oil, shale oil, tar sands, etc.)
5.Kinetic and Thermochemical Studies of In-Situ Chemical Processes
6.New Equipment for Testing and Field Application of Thermal EOR
7.In-situ Combustion (wet and dry combustion)
8.Laboratory Tools for Studying of Thermal EOR
9.High Pressure Air Injection (HPAI)
10.Ecological Impact of Thermal EOR
11.Cyclic Steam Stimulation (CSS)
12. Electromagnetic Heating EOR
13.EOR Field Applications
14.Steam Flooding
15.Flue Gas EOR

Peer review statement

All papers published in this volume of *IOP Conference Series: Earth and Environmental Science* have been peer reviewed through processes administered by the proceedings Editors. Reviews were conducted by expert referees to the professional and scientific standards expected of a proceedings journal published by IOP Publishing.

IOP Conf. Series: Earth and Environmental Science **155** (2018) 012001 doi:10.1088/1755-1315/155/1/012001

An analytical model of SAGD process considering the effect of threshold pressure gradient

P. Morozov[1], **A. Abdullin**[1,2], **M. Khairullin**[1]
[1]Institute of Mechanics and Engineering, Kazan Science Center, Russia
[2]Kazan Federal University, Kazan, Russia

E-mail: morozov@imm.knc.ru

Abstract. An analytical model is proposed for the development of super-viscous oil deposits by the method of steam-assisted gravity drainage, taking into account the nonlinear filtration law with the limiting gradient. The influence of non-Newtonian properties of oil on the productivity of a horizontal well and the cumulative steam-oil ratio are studied. Verification of the proposed model based on the results of physical modeling of the SAGD process was carried out.

1. Introduction

Steam-assisted gravity drainage (SAGD) is an efficient method for high-viscosity oil and natural-bitumen recovery based on steam injection and horizontal well technology [1]. Currently SAGD is used on heavy oil deposits in Canada, Venezuela and China. In Russia, the SAGD has been successfully applied on the Ashal'chinskoe field in the Republic of Tatarstan (since 2006) and the Yaregskoe field in the Republic of Komi.

The SAGD method uses a series of pairs of injection-producing horizontal wells (Figure 1). The steam chambers formed above each pair of wells, reaching the top of the formation, propagate horizontally until they coalescence [1, 2]. As the angle of inclination of the boundary of the steam chamber decreases, the rate of drainage decreases. At the final stage of the steam assisted gravity drainage in the inter-wellbore space stagnant zones are formed, not covered by the impact. The formation of such zones is a consequence of the manifestation of non-Newtonian properties of super viscous oils.

The efficiency of a SAGD project depends strongly on bitumen-production rate, recovery factor, and cumulative steam-oil ratio (CSOR). Hence, an accurate CSOR prediction is the key to the success of a SAGD project, particularly for the planning and engineering design phases.

Butler [1, 2] and Reis [3] assume that the shape of the steam chamber in a plane perpendicular to the wells is close to a triangle whose vertex coincides with the producing horizontal well. An analytical model of steam-assisted gravity drainage was proposed in [5], which describes the main stages of the evolution of the steam chamber: its growth up to the top of the formation, horizontal propagattion and expansion of the steam chamber in the direction of the bottom of the formation. In these works, the horizontal well production rate was calculated taking into account the Darcy law.

Unlike conventional oil, heavy oil and bitumen can be considered as a non-Newtonian (Bingham) fluid which has a threshold pressure gradient. However, the current analytical models as well as numerical simulation of SAGD process neglect the non-Newtonian flow behavior of heavy oil. In this work, a new analytical model based on the Butler's SAGD theory and filtration law with a threshold pressure gradient is developed and analyzed. This model allows to predict the oil flow rate and CSOR during all periods of steam chamber's growing.

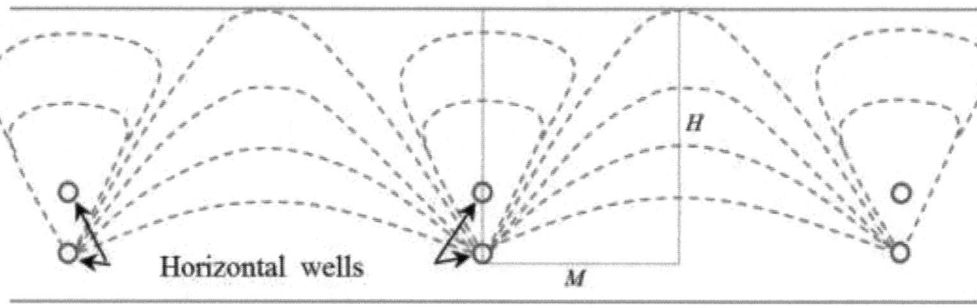

Fig. 1. SAGD scheme.

2. Analytical solution

It is assumed that oil is drained due to gravity according to a nonlinear filtration law with a threshold pressure gradient:

$$w = \begin{cases} \frac{k}{\mu}\left[\rho_0 g \sin\theta - \gamma\right], & \left|\rho_0 g \sin\theta\right| \ge \gamma, \\ 0, & \left|\rho_0 g \sin\theta\right| < \gamma, \end{cases} \tag{1}$$

where w – is the filtration velocity; ρ_0 – density of oil; θ – the angle of inclination of the boundary of the steam chamber; k – formation permeability; γ – characteristic value of the threshold pressure gradient for oil behind the contact surface of the steam chamber. The non-Newtonian nature of the flow of super-viscous oils is due to the high content of asphaltenes and high-molecular paraffins. One of the peculiarities of super-viscous oils is that they begin to "move" only after reaching the limiting pressure gradient γ, necessary to overcome the shear stress threshold τ_0. It has been experimentally established that the dependence γ vs τ_0 is of the form [4, 6]:

$$\gamma = \alpha \frac{\tau_0}{\sqrt{k}},$$

where $\alpha = (162 \div 180) \cdot 10^{-4}$.

Based on the law of conservation of mass and the nonlinear filtration law with threshold pressure gradient (1), an analytical model of steam assisted gravity drainage (2) - (4) is obtained, which describes the main stages of the steam chamber evolution in the element of super-viscous oil deposit development:

1. During the period of growth of the steam chamber, the production rate of a horizontal well varies according to a power law:

$$q(t) = 2L\left(\frac{3}{8}\operatorname{tg}\theta^{*}\phi\Delta S_0\right)^{\frac{1}{3}}\left(B - \frac{C}{\sin\theta^{*}}\right)^{\frac{2}{3}} t^{\frac{1}{3}}, \tag{2}$$

where $B = \dfrac{k\alpha g}{a\nu_s m}$, $C = \dfrac{k\alpha\gamma}{a\nu_s\rho_0 m}$, H – effective thickness of the formation (Fig. 1, Fig. 3a); L – length of horizontal well; ΔS_0 – the difference between the initial and final oil saturation; ϕ – porosity; α – thermal diffusivity; ν_s – kinematic viscosity of oil at the temperature of injected steam; g – acceleration of gravity; m – dimensionless parameter; $a = 0.4$ – an empirical constant [4]. According to Butler's experimental data, the angle θ^{*} formed by the boundary of the steam chamber and the bottom of the formation remains unchanged during the period of growth of the steam chamber to the top of the reservoir and amounts to ≈ 60 degrees [2].

The height of the steam chamber during the growth period is given by:

$$h(t)=\left(\frac{3}{4}\text{tg}\,\theta^*\right)^{\frac{2}{3}}\left(\frac{2B}{\phi\Delta S_0}-\frac{2C}{\phi\Delta S_0\sin\,\theta^*}\right)^{\frac{1}{3}}t^{\frac{2}{3}}. \quad (3)$$

The time the steam chamber reaches the top of the formation is determined from equation (3). The width of the steam chamber will then be $W_s^*=\dfrac{H}{\text{tg}\,\theta^*}$.

When $\gamma=0$ expression (2) is analogous to Butler's formula for estimating the production rate of a horizontal well during the growth of the steam chamber to the top of the formation [2].

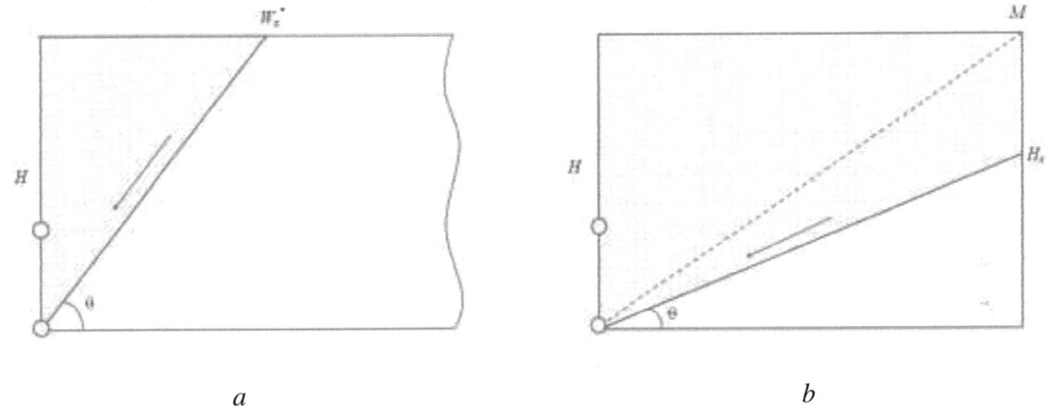

Fig. 2. Steam chamber scheme.

2. During the expansion of the steam chamber in horizontal direction, the production rate of a horizontal well is determinated from the solution of the ordinary differential equation with respect to the width of the steam chamber W_s:

$$q(t)=2LA\frac{dW_s}{dt},\ \frac{dW_s}{dt}=\left(\frac{B}{A}-\frac{C}{A\sin(\theta)}\right)^{\frac{1}{2}},\ W_s^*<W_s\le M\ , \quad (4)$$

where $A=\dfrac{\varphi\Delta S_0H}{2}$, $B=\dfrac{k\alpha g}{a\,v_sm}$, $C=\dfrac{k\alpha\gamma}{a\,v_s\rho_0m}$, $\theta=\text{arctg}\,\dfrac{H}{W_s}$, M – half the distance between neighbor pairs of horizontal wells (Figure 1, Figure 3 b). When $\gamma=0$ expression (4) coincides with the Reis formula [3]. In this case, the horizontal well rate during the horizontal expansion of the steam chamber remains constant.

3. During the period of expansion of the steam chamber in the direction of the bottom of the formation, the flow rate of the horizontal well is determined from the solution of the ordinary differential equation with respect to H_s:

$$q(t)=2LA\frac{dH_s}{dt},\ \frac{dH_s}{dt}=-\left(\frac{B\cdot H_s}{A}-\frac{C}{A\cos(\theta)}\right)^{\frac{1}{2}},\ H\ge H_s\ge H_s^*, \quad (5)$$

where $H-H_s$ – is the length of the common boundary of adjacent steam chambers (Figure 3b), $A=\dfrac{\varphi\Delta S_0M}{2}$, $B=\dfrac{k\alpha g}{a\,v_smM}$, $C=\dfrac{k\alpha\gamma}{a\,v_s\rho_0m}$, $\theta=\text{arctg}\,\dfrac{H_s}{M}$, $H_s^*=M\text{tg}(\theta_*)$ – height of the stagnant zone.

The limiting angle of inclination of the steam chamber boundary $\theta_* = \arcsin\left(\frac{\gamma}{\rho_0 g}\right)$. When the steam chamber reaches its limit position, the SAGD process is terminated. Note that when $\gamma = 0$ the flow rate of the horizontal well decreases linearly during this period of the steam chamber expansion[5].

The filtration law with the threshold pressure gradient (1) is an idealization of the filtration anomalies, and the value of γ is a dynamic characteristic, depending on the temperature. To take into account possible "subthreshold" flows, instead of the filtration law with the threshold pressure gradient (1), we also considered the piecewise linear law of filtration [6]:

$$w = \begin{cases} \frac{k(\theta)}{\mu}\left[\rho_0 g \sin\theta - \gamma\right], \left|\rho_0 g \sin\theta\right| \geq \frac{\gamma}{1-\varepsilon}, \\ \frac{\varepsilon k(\theta)}{\mu}\rho_0 g \sin\theta, \left|\rho_0 g \sin\theta\right| < \frac{\gamma}{1-\varepsilon}, \end{cases} \quad (6)$$

where $\frac{\varepsilon k}{\mu}$ – "subthreshold" mobility.

The main indicator of the effectiveness of the steam-thermal effect is the coefficient of the cumulative steam-oil ratio, which is defined as the ratio of the cumulative steam flow S to the cumulative oil production O [7]:

$$CSOR = \frac{S}{O}. \quad (7)$$

The cumulative steam flow S is calculated as the ratio of cumulative heat U_t to the latent heat of steam condensation U_l. In turn, U_t is the sum of the heat in the steam chamber and the total loss of heat through the top of the formation and lateral boundary of the steam chamber. The heat of the steam chamber is given by:

$$U_c = F\Delta T C_{vr},$$

where F – the area of the steam chamber (Figure 3), ΔT – the temperature difference between the steam chamber and the formation, C_{vr} – the heat capacity of the formation.

The loss of heat through the top of the reservoir is expressed in the form [7]:

$$U_o = \frac{8}{3} W_s \Delta T \sqrt{\frac{k_t C_{vo} t}{\pi}},$$

where k_t and C_{vo} – is the thermal conductivity and heat capacity of the top. The loss of heat through the lateral boundaries of the steam chamber is one third of the heat loss through the top of the formation [7].

3. Results and Discussions

Fig. 3 shows the mass flow of oil obtained in the course of the experiment on the physical model of the SAGD process [2, 8], as well as the results of flow rate calculations by the proposed model (solid line) with the following parameters: k = 2.5 μm^2, φ = 0.39, ΔS_0 = 0.95, H = 0.21 m, L = 0.03 m, M = 0.175 m, α = 0.05 m^2 / day, v_s = 9 m^2 / day, m = 3.6, γ = 0.001 MPa / m, ε = 0.05. The results of numerical calculations on the thermohydrodynamic simulator CMG STARS [8] are shown in Fig. 3 by the dotted line.

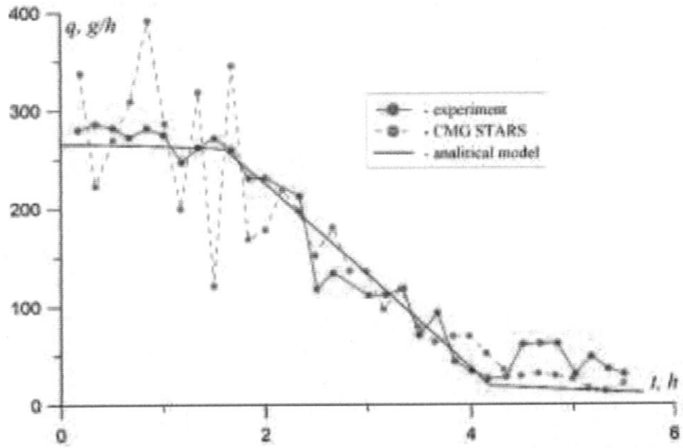

Fig. 3. Mass oil consumption by the proposed model, numerical model CMG STARS [8] and experimental data.

Fig. 4 shows the volumetric oil flow rate obtained during the experiment on the physical model of the SAGD process at low pressure of steam injection [9], as well as the results of production rate calculations by the proposed model with the following parameters: k = 240 μm², $\varphi = 0.35$, $\Delta S_0 = 0.9$, $H = 0.25$ m, $L = 0.08$ m, $M = 0.3$ m, $\alpha = 0.021$ m² / day, v_s = 3.8 m² / day, $m = 3$, γ = 0.001 MPa / m, $\varepsilon = 0.1$.

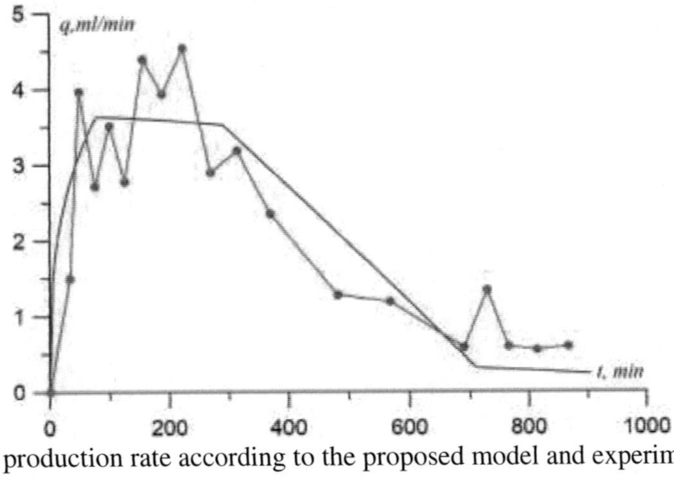

Fig. 4 Volumetric oil production rate according to the proposed model and experimental data [9].

Next, the reservoir model is considered with parameters [4]: $k = 2.5$ μm², $\varphi = 0.3$, $\Delta S_0 = 0.47$, $H = 25$ m, $L = 400$ m, $M = 50$ m, $\alpha = 0.03$ m²/day, $v_s = 0.8$ m²/day, $m = 4$, $\varepsilon = 0.01$. The results of calculations of the horizontal well rate, the cumulative steam-oil ratio and cumulative oil production for different values of the threshold pressure gradient γ_0 are presented in Fig. 5. As shown in Fig. 5 the threshold pressure gradient exerts a significant influence on the dynamics of the main SAGD indicators at all stages of growth of the steam chamber.

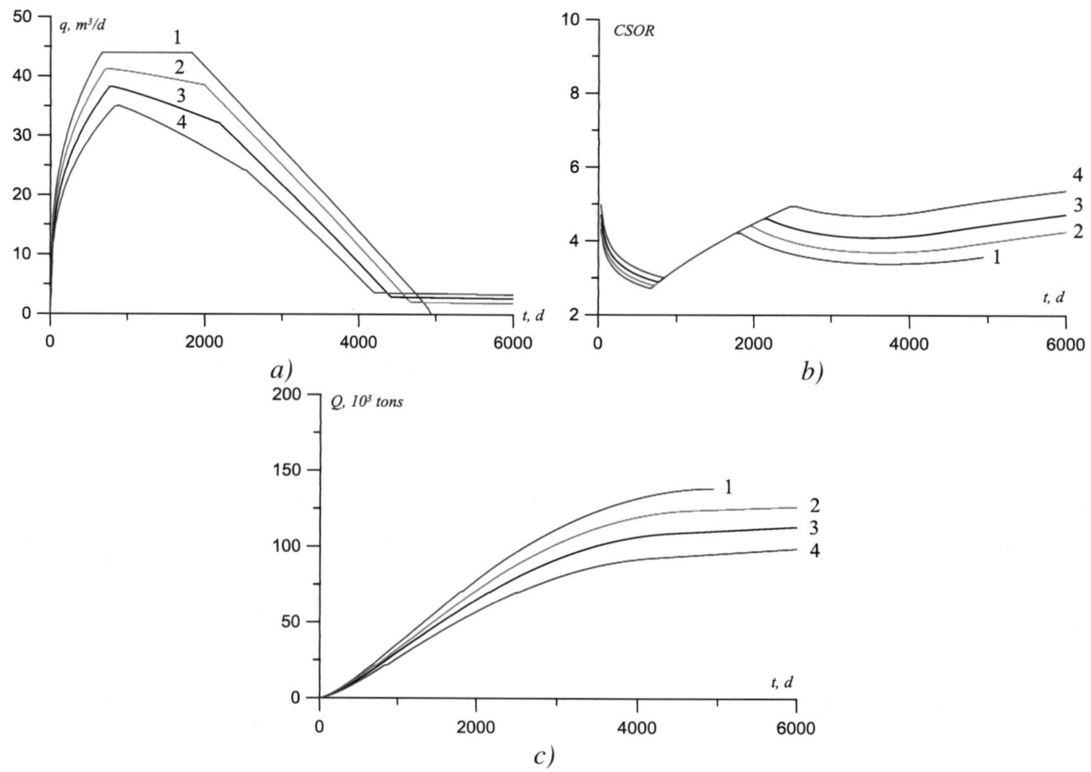

Fig. 5. Influence of the threshold pressure gradient on the horizontal well production rate (a), the cumulative steam oil ratio (b), cumulative oil production (c).
(1 - $\gamma = 0$ MPa, 2 - $\gamma = 0.001$ MPa, 3 - $\gamma = 0.002$ MPa, 4 - $\gamma = 0.003$ MPa, $\varepsilon = 0.01$)

The results indicate that the presence of threshold pressure gradient can significantly reduce the horizontal well productivity and increase the CSOR. Also, the threshold pressure gradient leads to a limiting angle of inclination of the steam chamber boundary and to the formation of an undrained zones between pairs of horizontal wells. It is shown that when the threshold pressure gradient is zero value, the proposed analytical model of SAGD process reduces to previous our model based on the Newtonian flow behavior of heavy oil [2].

4. Conclusions

An analytical model to predict the production rate of a horizontal well in SAGD process was developed by taking into account the effect of the threshold pressure gradient. The proposed model can predict the oil flow rate and CSOR during three periods of steam chamber's growing, and was verified against experimental data on physical models. It shown that the threshold pressure gradient can significantly reduce the horizontal well productivity and increase the CSOR.

References

[1] Butler R M 2010 *Horizontal wells for oil, gas and bitumen production* M - Izhevsk: Institute for Computer Research, SRC "Regular and chaotic dynamics" p 536 [in russian]

[2] Butler R 1991 *Thermal recovery of oil and bitumen* (New Jersey: Prentice Hall)

[3] Reis J C 1992 A steam-assisted gravity drainage model for tar sands: linear geometry *J Can Pet Tech* vol **31** №10 pp 14–20

[4] Khisamov R S , Musin M M , Musin K M , Fayzullin I N , Zaripov A T 2013 Generalization of the results of laboratory and pilot-industrial work on the extraction of super-viscous oil from the reservoir Kazan: *Publishing house "Fen" of the Academy of Sciences of the Republic of Tatarstan,* 232 p [in russian]

[5] Khisamov R , Morozov P , Khairullin M , Shamsiev M and Abdullin A 2015 *Oil Industry* vol **4** p 62

[6] Bernadiner M G , Entov V M 1975 *Hydrodynamic theory of the filtration of anomalous liquids* Moscow: Nauka, 199 p [in russian]

[7] Miura K , Wang J 2012 An analytical model to predict cumulative steam/oil ratio (CSOR) in thermal-recovery SAGD process *J Can Pet Tech* vol **51** №4 pp 268–275

[8] Chow L , Butler R M 1996 Numerical simulation of the steam-assisted gravity drainage process (SAGD) *J Can Pet Tech* vol **35** №6 pp 55–62

[9] Shijun H , Hao X , Shaolei W , Chenghui H , Yang Y 2016 Physical simulation of the interlayer effect on SAGD production in mackay river oil sands *Fuel* vol **183** №3 pp 373–385

Studying metal impurities (Mn^{2+}, Cu^{2+}, Fe^{3+}) in calcium phosphates by electron paramagnetic resonance

K. Iskhakova[1,*], F. Murzakhanov[1], G. Mamin[1], V. Putlyaev[2], E. Klimashina[2], I. Fadeeva[3], A. Fomin[3], S. Barinov[3], A. Maltsev[3], S. Bakhteev[4], R. Yusupov[4], M. Gafurov[1], and S. Orlinskii[1]

[1]Kazan Federal University, 18 Kremlevskaya Str., Kazan, Russia
[2]Lomonosov Moscow State University, GSP-1, Leninskie Gory, Moscow, Russia
[3]Baikov Institute of Metallurgy and Materials Science, 49 Leninskii Pr., Moscow, Russia
[4]Kazan National Research Technological University, 41 K. Marxa Str., Kazan, Russia

* E-mail: kamilaiskh@gmail.com

Abstract Calcium phosphates (CaP) are exploited in many fields of science, including geology, chemistry, biology and medicine due to their abundance in the nature and presence in the living organism. Various analytical and biochemical methods are used for controlling their chemical content, structure, morphology, etc. Unfortunately, magnetic resonance techniques are usually not even considered as necessary tools for CaP inspection. Some aspects of application of the commercially realized electron paramagnetic resonance (EPR) approaches for characterization of CaP powders and ceramics (including the nanosized materails) such as hydroxyapatite and tricalcium phosphates of biogenic and synthetic origins containing intrinsic impurities or intentional dopants are demonstrated. The key features and advantages of the EPR techniques for CaP based materials characterization that could compliment the data obtained with the recognized analytical methods are pointed out.

1. Introduction

Calcium phosphates (CaP) are interesting compounds in many fields of science, including geology, chemistry, biology and medicine due to their abundance in the nature and presence in the living organism. In nature, different calcium phosphate minerals are produced within a wide range of environments by geological (igneous apatite), geochemical and/or geomicrobiological (phosphorite), and biological (biological apatite) processes. Igneous apatite minerals nucleate and crystallize from molten, phosphate-rich rock, forming crystalline fluorapatite ($Ca_5[PO_4]_3F$), chlorapatite ($Ca_5[PO_4]_3Cl$), or hydroxyapatite ($Ca_{10}[PO_4]_6[OH]_2$, HA or HAp) [1-4].

Manganese, copper and iron are essential impurity elements in CaP. Oil containing formation – rock, sands or bitumen very often possess impurity manganese and iron [5]. It is worth noting that quantitative determination of copper, manganese and iron in calcified materials is still a challenge for the analytical tools because of the structural and chemical complexity of the investigated matrix. One of the established experimental methods for identification of these ions in different types of materials (including apatites) and characterization of structure of their complexes is electron paramagnetic resonance (EPR) [6-23].

CaP materials with different functional features such as biocompatibility, catalytic activity, ion conductivity may be formed by the cation or/and anion substitution within initial matrix. For example, the experimental results indicate that HAp/MnO_2 composite may be an effective adsorbent for the removal of lead ions from aqueous solutions [24], while the HAp supported manganese could serve as precursor for the oxide catalysts [25]. One more characteristic was established recently for the lanthanide modified CaP – pigmentation [26, 27], that make this substances promising to be used for the development of bioactive materials for bone regeneration and smart fluorescent probes for bio imaging applications.

A huge number of publications were and are devoted to the cationic and anionic substitutions in the HAp structure while the second most frequent CaP - tricalcium phosphate (TCP) mainly in its β-form (β-TCP), is studied much less then HAp (see [28] and references therein). Despite the large number of the studies performed, many important problems relating to anionic and cationic substitutions in HAp are not sufficiently investigated. Moreover, data of different studies are contradictive. The most contradictive information is associated with the sites of the ions localization in biomineral, synthetic and nanosized samples.

Though even conventional X-band (microwave frequency of about 9 GHz) EPR is a sensitive tool for the CaP characterization, it is still not widely applied for these purpose and mainly restricted by the investigation of the radiation induced oxygen and carbonate radicals. One of the aims of the manuscript is to present some useful applications of the conventional X-band EPR and to show some abilities of the "unconventional" EPR techniques to encourage the CaP specialists to use different EPR approaches.

2. Materials and methods

Pure HAp, TCP and their substituted powders were obtained using a precipitation technique [29] according to the following reaction:

$$(3\text{-}x)Ca(NO_3)_2 + xM(NO_3)_2 + 2(NH_4)_2HPO_4 \rightarrow Ca_{3-x}M_x(PO_4)_2, \qquad (1)$$

where M= Cu^{2+}, Fe^{3+}, Mn^{2+}, x= 0;0,01; 0.05; 0.1 or 1.0. (For HAp the reaction scheme was the similar).

Synthesis procedure was as follows: the 0.5 mol/L solution of $Ca(NO_3)_2$ were mixed with a calculated amount of metal nitrate solution; then $(NH_4)_2HPO_4$ solution was added dropwise with the rate of 20 mL/min. The pH level was kept at 6.5-6.9 by the dropwise addition of ammonia solution. After 30 min, the precipitate was filtered, washed with distilled water and dried at 80°C. The precipitate was calcined at 900°C in order to remove the traces of NH_4NO_3 and to form the whitlockite structure.

The samples were characterized by infrared spectroscopy (Nikolet Avatar 330), atomic emission spectroscopy with inductively coupled plasma (Bruker Ultima-2), X-ray difractometry (Shimadzu 6000, Bruker D8 Advance, Rigaku D/Max-2500), X-ray fluorescence analysis with total external reflection (S2 Picofox, Bruker).

The biogenic CaP containing samples were obtained during carotid endarterectomy in patients with atherosclerosis (the inner lining of the carotid artery specimens). We investigated 27 endarterectomy specimens (plaques or fragments of plaques) derived from 27 male patients with a mean age of 57 years (42 to 64) of whom 19 were symptomatic. Based on the modified Gray-Weale classification of atherosclerotic plaques, 21 plaques were classified as "unstable" (echolucent, low median gray level, GSM $\leq$ 25) while 6 as "stable" (echogenic, GSM > 25). For EPR studies the tissue specimens were transferred and kept in liquid nitrogen (T = 77 K), then freeze-dried (lyophilized) at $P = 5\cdot10^{-5}$mBar and T = 253 K (- 20°C). More details are given in paper [16].

EPR measurements were done by using W-band (with the microwave frequency of about 93.5 GHz) Bruker Elexsys 680 spectrometer and X-band (frequency of about 9.5 GHz) Bruker Elexsys 580 and table-top Labrador (Ekaterinburg, Russia).

IOP Conf. Series: Earth and Environmental Science **155** (2018) 012002 doi:10.1088/1755-1315/155/1/012002

3. Results and Discussion

3.1. Conventional EPR

Pure HA and TCP are EPR silent. Consequently, EPR can be used for the purity check and quantitative analysis of the doped CaP materials. One of the EPR features (in addition to the fact that it is a non-destructive method) is the high selective sensitivity to the complexes of the paramagnetic complexes of Mn, Fe and Cu (Figures 1, 2, 3). Though the increase of PC amount is followed by the raise of signal intensity, due to the electron spin-spin interaction the high concentrations of dopants could lead to the broadening of EPR lines and hinders the unambiguous interpretation of EPR spectra. For example, EPR pattern of HA+1 % Mn^{2+} resembles that for 0.01 % of iron (cf. Figures 1 and 2).

Another remarkable feature is the sensitivity of EPR parameters to the interaction of PC with the magnetic moments of the own and neighbouring nuclei (hyperfine and super-hyperfine interaction). Because the nuclear spin I = 5/2 for ^{55}Mn, six lines (2I + 1 = 6) of hyperfine (HF) structure for Mn^{2+} with the nearly isotropic HF constant $A_{iso} \approx$ (9-10) mT is to obtain (Figures 1 and 4).

In case of even slight anisotropy of magnetic interaction tensors (such as g- and hyperfine A tensors), EPR can provide the information about their orientation dependencies and thereby about the spatial structure of the paramagnetic complexes in disordered systems. For copper both stable isotopes ^{63}Cu and ^{65}Cu have I = 3/2 and close magnetic moments. This and the known anisotropies of the spectroscopic parameters g and A for Cu^{2+} in different matrices directly allow to assign presented in Figure 3 spectra to Cu^{2+} due to 4 lines detected at $g_z \approx 2.4$ with $A_z \approx$ 200-250MHz.

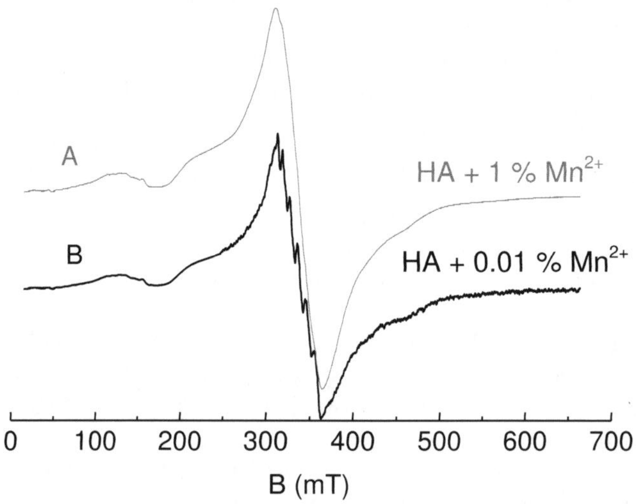

Figure 1. Comparison of EPR spectra of the manganese doped HAp at 9 GHz and T = 300 K for (A) 1% and (B) 0.01 % concentrations. Six lines of the hyperfine structure due to the I = 5/2 for ^{55}Mn nuclei is to observe for 0.01% Mn^{2+}. The signal amplitude (vertical axis) for 1 % Mn^{2+} is reduced in 100 times.

IOP Conf. Series: Earth and Environmental Science **155** (2018) 012002 doi:10.1088/1755-1315/155/1/012002

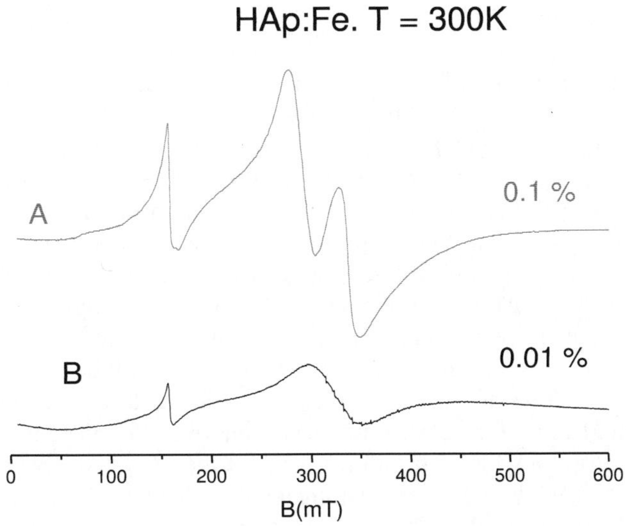

Figure 2. Comparison of of EPR spectra of Fe doped HA at 9 GHz and T = 300 K for (A) 0.1% and (B) 0.01% concentrations. Signals et (100-160) mT indicate that iron at least partially is in Fe^{3+} state.

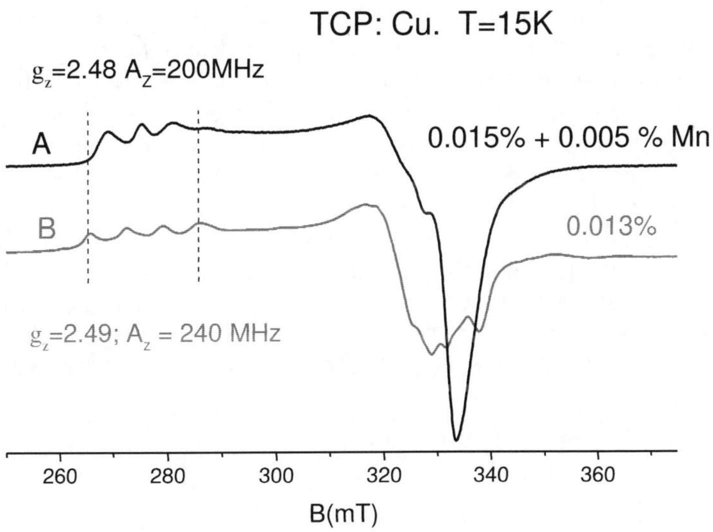

Figure 3. Comparison of EPR spectra of Cu doped β-TCP at 9 GHz and T = 15 K. During the synthesis, one sample (upper curve A) was co-doped with a small amount of manganese that slightly changed the EPR pattern, values of g- components and hyperfine (A_z) splitting compared to the "pure" Cu-doped TCP (curve B). The concentration of paramagnetic centers was derived after the double integration of EPR spectra compared to the reference one with the known concentrations. Vertical dashed lines are drawn to show the difference visually.

IOP Conf. Series: Earth and Environmental Science **155** (2018) 012002 doi:10.1088/1755-1315/155/1/012002

3.2. High-field and pulse EPR

Sometimes observation and interpretation of the HF structure (HFS) and superhyperfine structure (SHS) is a rather complicated task. Even in the perfect crystals the resolution of the SHS spectra, the number of the components of the structure, their relative intensities strongly depend on the concentration of PC, orientation of the applied magnetic field and the EPR frequency [30]. That is why, for example, the resolved HFS of Mn^{2+} or HFS for Cu^{2+} ions in the (x, y) plane (corresponding to the perpendicular orientation of B to the main axes of g-tensor) is rarely observed at X-band in the disordered systems (powders, ceramics, nanosized samples, inhomogeneous samples of biogenic origin, samples containing non-crystal amorphous phase, etc) and a transition to the higher frequencies are necessary to interpret the EPR spectra correctly (Figure 4).

Pulse methods are able to significantly increase and supplement information obtained from conventional EPR spectra. While stationary method uses continuous microwave irradiation and reveals splitting of energy levels of paramagnetic system, pulse techniques provide insights into the dynamics of the system and allows to measure relaxation times, decode complex EPR spectra of several interacting PC, etc. Because the electronic relaxation times (longitudinal or spin-lattice relaxation time T_{1e} and transverse or spin-spin relaxation time T_{2e}) are of 3-6 orders of magnitude shorter than the nuclear ones, the characteristic time parameters are also much shorter than in nuclear magnetic resonance (NMR) experiments. The connected with those technical difficulties of creating short powerful high frequency pulses prevented till recently wide applications of the pulsed EPR approaches. The typical pulse sequences we use in our research in the W-band (94 GHz) are: (1) $\pi/2 - \tau - \pi$ with the $\pi/2$ pulse duration of 32 ns and the time delay $\tau = 240$ ns to obtain electron spin echo (ESE, Figure 4); (2) T_{2e} can be studied by tracking the primary ESE amplitude with the same $\pi/2$–π pulse durations while varying τ (Figure 5); (3) T_{1e} can be extracted from the inversion-recovery studies by applying the $\pi - T_{delay} - \pi/2 - \tau - \pi$ pulse sequence, while T_{delay} is varied.

From Figure 4 it follows that in the atherosclerotic plaques small amount (less than 0.005 %) of Mn^{2+} is obtained most probably in the crystal phase of the nanosized HAp constituent. EPR spectra and relaxation characteristics (Figure 5) differ for stable and unstable plaques that can be connected with the more inhomogeneous distribution of Mn^{2+} ions in the HA structure [16]. Consequently, EPR opens one more dimension in the investigations of the calcification processes.

Since recently the technique of the simultaneously co-doping of CaP by various ions and groups to achieve desired properties attracts a great attention [31, 32]. Spectral and relaxation characteristics of PC are very sensitive to the changes in the local environment and, therefore, could be used for studying co-doping effects even if the EPR spectrum of the second (or the third) codopant is not detected because of the low concentration or masking under the EPR spectra of other PC. Figure 3 illustrates that HFS of Cu^{2+} ions in TCP changes with Mn^{2+} addition though the Mn^{2+} spectrum is blinded by Cu^{2+} one. In papers [13, 14] we have observed and investigated the influence of Mn^{2+} on the relaxation parameters (T_{2e}) of the impurity NO_3^-/NO_3^{2-} group in the micro and nano-sized HAp.

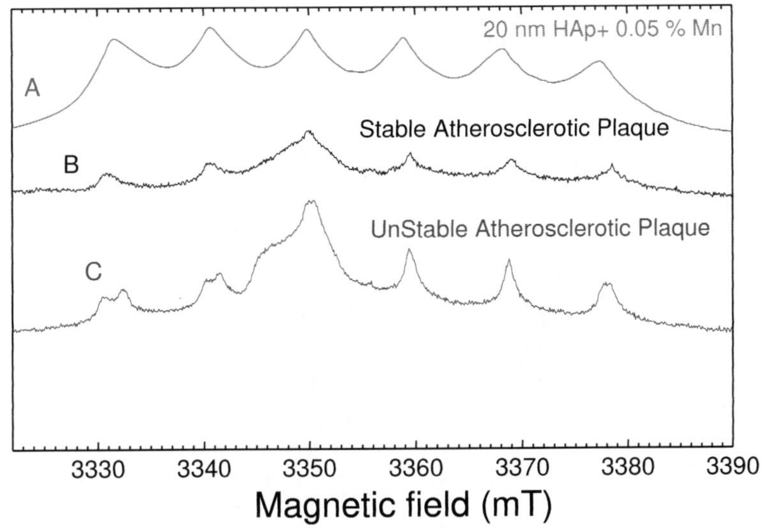

Figure 4. Comparison of pulsed EPR spectra at 94 GHz and T = 50 K of (A) synthetic nanohydroxyapatite with 0.05 wt % of Mn^{2+}; (B) stable and (C) unstable atherosclerotic plaques obtained during carotid endarterectomy in patients with atherosclerosis. Due to the small amount of samples and their inhomogeneity, the manganese spectra could not be obtained at conventional for EPR X-band (9 GHz) frequencies.

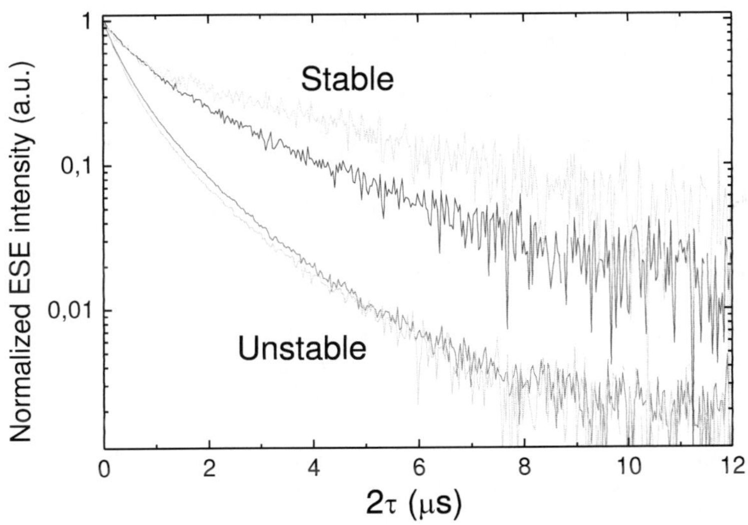

Figure 5. ESE decay (T_{2e} curves) for Mn^{2+} ions in stable (two upper curves) and unstable (two lower curves) atherosclerotic plaques taken at W-band and T = 8 K.

4. Conclusion

The modern commercially realized EPR techniques offer new options for the rapid testing and comprehensive examination of the condensed matter. In our opinion, the possibilities of even conventional EPR approaches for investigation and analysis of CaP materials are still not fully exploited and described in details. We hope that our paper would encourage the specialists in different scientific and industrial branches to use different EPR approaches in their studies and routine analyses.

Acknowledgments
The authors are thankful to T. B. Biktagirov (Kazan Federal University) for theoretical calculations. The work is financially supported by Russian Foundation for Basic Research under Grant #18-32-00337. E.K and V.P. acknowledge partial support from Lomonosov Moscow State University Program of Development,

References
[1] Barinov S and Komlev V 2008 Calcium phosphate based bioceramics for bone tissue enginery, , (Zurich: Trans Tech. Publ.)
[2] Omelon S, Ariganello M, Bonucci E, Grynpas M and Nanci A 2013 *Calcif. Tissue* Int. **93** 382
[3] Barinov S 2010 *Russ. Chem. Rev.* **79** 13
[4] Uskovic V and Vu V 2016 *Materials* **9** 434
[5] Khasanova N M, Gabdrakhmanov D T, Kayukova G P, Mikhaylova F N and Morozov V P 2017 *Magn. Reson. Solids* **19** 17102
[6] Low J 1957 *Phys. Rev.* **105** 793
[7] Gilinskaya L G and Sherbakova M Y 1971 *J. Struct. Chem.* **11** 950
[8] Gilinskaya L G and Sherbakova M Y 1975 *Isomorphous substitution and structural defects in apatites as found by EPR investigation* (Novosibirsk: Publishing House "Nauka") p 62 (in Russian)
[9] Mayer I, Diab H, Reinen D and Albrecht C 1993 *J. Mater. Sci.* **28** 2428
[10] Mayer I, Jacobsohn O, Niazov T, Werckmann J, Iliescu M, Richard-Plouet M, Burghaus O and Reinen D 2003 *Eur. J. Inorg. Chem.* **7** 1445
[11] Pan Y, Chen N, Weil J A and Nilges M J 2002 *Amer. Miner.* **87** 1333
[12] Pon-On W, Meejoo S and Tang I 2008 *Mater. Res. Bull.* **43** 2137
[13] Gafurov M, Biktagirov T, Mamin G, Klimashina E, Putlayev V, Kuznetsova L and Orlinskii S 2015 *Phys. Chem. Chem. Phys.* **17** 20331
[14] Gafurov M R, Biktagirov T B, Mamin G V, Shurtakova D V, Klimashina E S, Putlyaev V I and Orlinskii S B 2016 *Phys. Solid State* **58** 469
[15] Stich T A, Lahiri S, Yeagle G, Dicus M, Brynda M, Gunn A, Aznar C, DeRose V J and Britt R D 2007 *Appl. Magn. Reson.* **31** 321
[16] Chelyshev Y, Gafurov M and Ignatyev I 2016 *BioMed Res. Int.* **2016** 3706280
[17] Galukhin A V, Khelkhal M A, Gerasimov A V, Biktagirov T B, Gafurov M R, Rodionov A A and Orlinskii S B 2016 *Energy Fuels* **30** 7731
[18] Fadeeva I, Gafurov M and Kiiaeva I. 2017 *BioNanoScience* **7** 434
[19] Gafurov M, Biktagirov T, Mamin G and Orlinskii S 2014 *Appl. Magn. Reson* 45 1189
[20] Gafurov M, Biktagirov T and Yavkin B 2014 *JETP Lett.* **99** 196
[21] Biktagirov T, Gafurov M, Mamin G, Klimashina E, Putlayev V, Orlinskii S 2014 *J. Phys. Chem. A* **118** 1519
[22] Yavkin B, Mamin G and Orlinskii S. 2012 *Phys Chem Chem Phys* **14** 2246
[23] Biktagirov T, Gafurov M, Iskhakova K, Mamin G and Orlinskii S 2016 *J. Low Temp.Phys.* **185** 127
[24] Dong L, Zhu Z, Qiu Y and Zhao J 2016 *Front. Environ. Sci. Eng.* **10** 28
[25] Chlala D, Giraudon J M, Nuns N, Lancelot C, Vannier R N, Labaki M and Lamonier J F 2016 *Appl. Catal. B Environ.* **184** 87
[26] Pogosova M, Eliseev A, Kazin P and Azarmi F 2017 *Dyes and Pigments* **141** 209
[27] Kaura K, Singh K, Anand V, Islam N, Bhatia G, Kalia and Singh J 2017 *Ceram. Int.* **43** 10097
[28] Zilm M, Chen L, Sharma V, McDannald A, Jain M, Ramprasad R and Wei M 2016 *Phys. Chem. Chem. Phys.* **18** 16457
[29] Fadeeva I, Selezneva I and Davydova G 2016 *Doklady Chemistry* **468** 159
[30] Aminov L, Gafurov M, Korableva S, Kurkin I and Rodionov A 2017 *Phys.Solid State* **59** 564
[31] Kulanthaivel S, Mishra U and Agarwal T 2015 *Ceram. Int.* **41** 11323
[32] Singh R, Srivastava M, Prasad M and Kannan S 2017 J. Alloys Compd.**725** 393

Mims electron-nuclear double resonance in $LiYF_4:Ce^{3+}$ crystal

M Gafurov, G Mamin, I Kurkin, and S Orlinskii

Kazan Federal University, 18 Kremlevskaya Str., Kazan 420008, Russia

e-mail: Marat.Gafurov@kpfu.ru

Abstract. We report the observation of the pulsed electron-nuclear double resonance (ENDOR) spectra from ^{19}F and ^{7}Li nuclei on impurity Ce^{3+} ions in $LiYF_4$ crystal. The resolved structure from the nearby and remote nuclei in spectra is observed. The outcome shows that $LiYF_4:Ce^{3+}$ system can be exploited as a convenient matrix for performing spin manipulations and adjusting quantum computation protocols while ENDOR technique is usable for the investigation of electron-nuclear interaction with all the nuclei of the system.

1. Introduction

As a significant class of rare earth compounds, rare earth (RE) fluorides have been become anew a research focus in the material field due to their unique applications in optical communications, three-dimensional displays, solid-state laser, catalysis, solar cells, biochemical probes and medical diagnostics [1], [2]. It is proposed that RE doped fluoride glasses as parts of laser systems could be fruitfully used for determining identity and/or quantity of a component of a fluid at a remote location such as downhole in a wellbore or inside a pipeline, particularly at elevated temperatures up to 175° C, such as an oil based mud or crude oil, and may detect components such as methane, ethane, carbon dioxide, hydrogen sulfide, etc. [3]. The potential for using RE doped crystals as scintillators (detectors) for radioactive radiation is not disclosed [4]. To make good use of the excellent properties of metal fluorides, their morphology and microstructure should be carefully controlled. One of the powerful method of the RE doped crystals inspection is an electron paramagnetic/spin resonance (EPR/ESR).

Apart from being the cornerstone of modern laser technology, crystals and glasses doped with RE metals are known to have excellent properties for classical and quantum information storage and for performing quantum computation protocols [5], [6], [7]. In addition, $LiYF_4$ crystals, having a relatively simple crystalline structure and growing almost without defects, are unique model systems for studying the crystal field, electron-phonon, interionic, hyperfine interactions, and also isotopic effects [8].

The interaction of paramagnetic ions in crystals with the magnetic moments of the ligand nuclei (super-hyperfine interaction) usually leads to broadening of the EPR lines, their overlap and even the impossibility of observing the allowed super-hyperfine structure (SHS) in the spectra [8]. Analysis of the SHS allows one to obtain important information on inter-ion interactions and distribution of electron spin density in crystals. For fluorides SHS can vary in a huge range even in the same ligand environment depending on the paramagnetic ion: from 10 MHz for RE ions up to 300 MHz for Pb^{3+} [9], [10]. The nuclear Zeeman splitting for F^- is also of about 10 MHz that complicates the decoding of the EPR spectra.

To the best of our knowledge, up to now SHS with only the nearest fluorine environment (first coordination sphere) has been experimentally and theoretically investigated. Here we show that exploiting pulsed electron-nuclear double resonance (ENDOR) techniques one can study the super-hyperfine interaction with the remote nuclei.

2. Experiment and Discussion

Crystals of $LiYF_4$ doped with ≈ 0.05 at % of Ce^{3+} were grown in Kazan Federal University by the Bridgman-Stockbarger technique in argon atmosphere. EPR investigations were done by using helium flow cryostats at X-band Bruker ESP 300 и Bruker Elexsys E580 spectrometers (T = 5-25 K, microwave frequency ν ≈ 9.5 GHz). At these temperatures an intense EPR spectrum from Ce^{3+} ions was observed (Figure 1). Resonant magnetic fields, B, in the orientations B||c and B⊥c (c is the optical axis of crystal) correspond to the g-factors given in paper [11] : $g_{\parallel} = 2.737$; $g_{\perp} = 1.475$. For Ce^{3+} ions none of the stable isotopes (^{136}Ce, ^{138}Ce, ^{140}Ce, ^{142}Ce) has a nuclear spin and, therefore, no intrinsic hyperfine structure (HFS) exists.

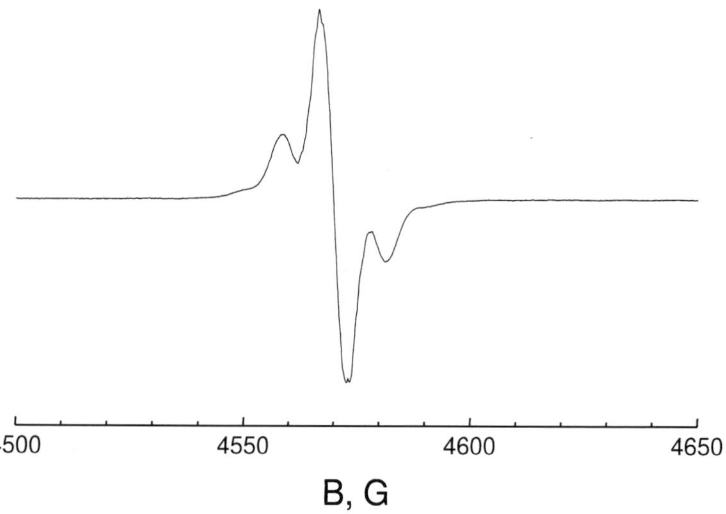

Figure 1. EPR line with the SHS of Ce^{3+} ions in $LiYF_4$. **B**⊥c, T=20K, ν = 9.38GHz.

We used two–pulse electron spin echo measurements for the primary Hahn–echo sequence ($\pi/2$–τ–π–τ–echo), where τ is the interpulse delay time of 240 ns, with initial $\pi/2$ and π pulse lengths of 18 and 36 ns, respectively.

For ENDOR experiments we used special double (for nuclei and electron) cavities and Mims pulse sequence - $\pi/2$-τ-$\pi/2$-T-$\pi/2$ with an additional radiofrequency (RF) pulse π_{RF} = 18 μs inserted between the second and third microwave $\pi/2$ pulses (see Figure 2). RF frequency in our setup could be swept in the range of (1-200) MHz. Magnetic field value was kept constant, corresponding to the orientation of B⊥c. Some ENDOR basics are briefly described below. Some examples of the studying petroleum asphaltenes and crude oils by ENDOR technique are presented in our recent papers [12], [13]

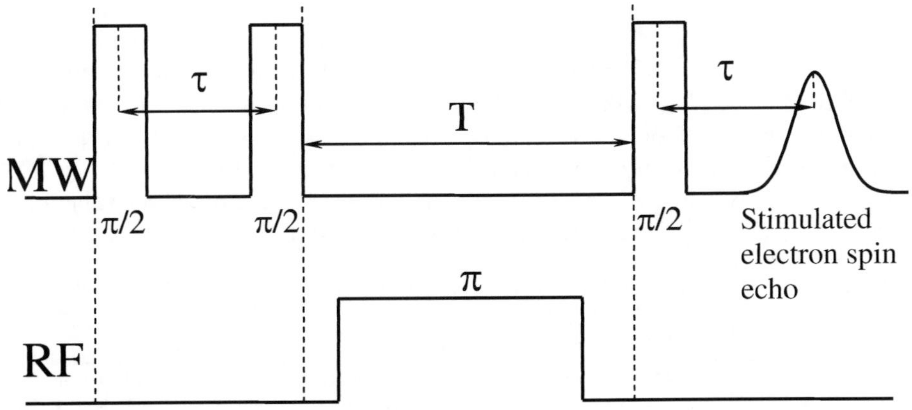

Figure 2. Mims pulse sequence at microwave (MW) and radiofrequencies (RF) used to obtain the ENDOR spectra as a function of stimulated electron spin echo amplitude from the frequency of RF pulse.

In the case of a "free" nucleus, RF pulse applied at the Larmor frequency

$$\nu_{Larmor} = |\gamma B_0| \equiv h^{-1}|g_I \beta_I B_0|, \quad (1)$$

where γ is a gyromagnetic ratio of the nuclear spin *I, h* is a Planck constant, g_I is a nuclear g-factor and β_I is a nuclear Bohr magneton, can change the state of the nuclear spin (the population of the nuclear sublevels). For the hyperfine coupling constant *A* and simple electron-nuclear coupling ($S = 1/2, I = 1/2$), it can lead to the appearance of the characteristic features in the ENDOR spectrum at the RF frequencies

$$\nu_{ENDOR} = \nu_{Larmor} \pm A/2 \text{ or} \quad (2)$$

$$\nu_{ENDOR} = A/2 \pm \nu_{Larmor}, \quad (3)$$

depending on the ratio between *A* and ν_{Larmor}.

The ENDOR splitting, a_{ENDOR}, can help not only to identify a type of nuclei coupled with the electron spins but also provide spatial relationships between them. For the pure electron-nuclei dipole-dipole interaction in the point model, the electron-nuclear distance, *r*, from the ENDOR splitting can be estimated from

$$a_{ENDOR} \approx g \cdot g_I \cdot (1 - 3\cos^2\theta)/r^3, \quad (4)$$

where g is a g-factor of electron spin *S*, Θ is an angle between directions of the parallel component of g ($g_{||}$) and B_0. As it follows from (4), a_{ENDOR} depends on the distance between the electron and nuclear spins and their mutual orientation. Additionally, for $I = 1$ an electric nuclear quadrupole coupling exists that can split or shift the ENDOR lines: the nuclear quadrupole interaction is sensitive to the electric field gradient at the site of the nucleus.

All nuclei in $LiYF_4$ compound have sensible magnetic moment capable to create a perceptible additional magnetic field at the location of the rare earth impurity. In Table 1 some data for them which are important for the interpretation of ENDOR spectra are presented.

ENDOR spectra for B⊥c presented in Figure 3 are clearly due to the electron-nuclear interactions with ^7Li and ^{19}F isotopes. The ability to observe narrow ENDOR lines at various RF frequencies from the Larmor ones testifies that Ce^{3+} is an appropriate probe to sense remote nuclei (see Eq. 4).

Interpretation of ^{7}Li ENDOR causes troubles due to the possible influence of quadrupole interaction (see Table 1). From ENDOR pattern for ^{19}F one can estimate the electron-nuclear distance to the most remote nuclei from Eq. (4) as $r \approx 0.66$ nm. Taking into account that in the YF_8 octahedra of $LiYF_4$ structure, the four first F neighbours are situated at 0.224-0.226 nm ($a_{ENDOR} \approx 4.2$ MHz), the distance to the four second neighbours is of about 0.229-0.232 nm [8], one can see that the electron spin density of Ce^{3+} ions distributes well out of even the second coordination sphere. Technical limitations did not allow us to register low frequency (of about 1 MHz) ENDOR features.

Table 1. Nuclear spin, natural abundance, relative gyromagnetic ratio and calculated Larmor frequency in B = 4720 G for the $LiYF_4$ nuclei

Nuclei	Isotope	I	Natural Abundance, %	Relative to ^{1}H gyromagnetic ratio	Larmor frequency for B = 4720 G, MHz
3Li	6	1	7.6	0.1471	2.96
	7	3/2	92.4	0.3887	7.81
9F	19	1/2	100	0.9412	18.91
39Y	89	1/2	100	0.0492	0.99

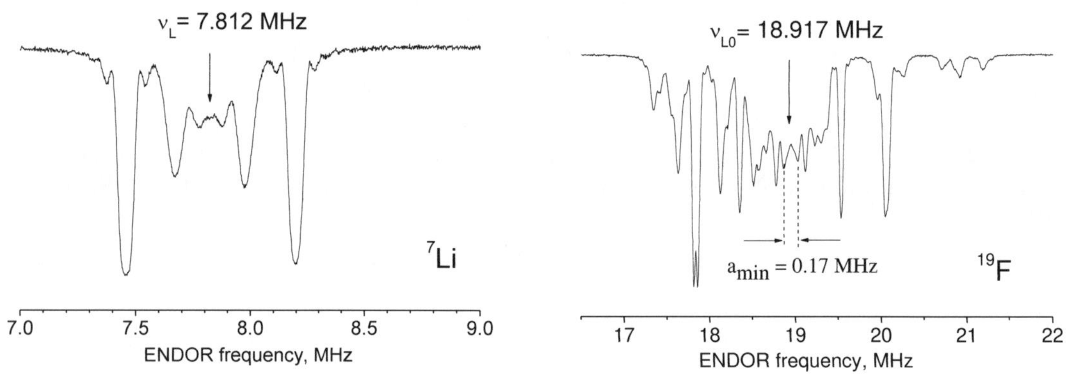

Figure 3. ENDOR spectra caused by interactions with ^{7}Li (left panel) and ^{19}F (right panel) isotopes in $LiYF_4$: Ce^{3+} monocrystals in B⊥c orientation. B= 4721 G, T=10 K, ν = 9.75 GHz. Corresponding Larmor frequencies are marked by arrows. Dashed lines mark the minimal observed ENDOR splitting for ^{19}F.

The observed features show that $LiYF_4$:Ce^{3+} system potentially can be used as an element of quantum computing. Indeed, it is known that electrons are natural candidates as physical qubits to be exploited for quantum computing and information processing (QC/QIP), and therefore magnetic resonance techniques, consisting of EPR and NMR (nuclear magnetic resonance) are among the most appropriate techniques to be exploited for quantum computing. Pulsed EPR enables manipulation of electron spin and nuclear spin qubits in an equivalent manner and ENDOR can be used as the most useful spin manipulation technology in implementing QCs/QIPSs [14].

Acknowledgments
The work is financially supported by the Russian Science Foundation, Project # 17-72-20053.

IOP Conf. Series: Earth and Environmental Science **155** (2018) 012003 doi:10.1088/1755-1315/155/1/012003

References

[1] Tressaud A 2010 *Functionalized Inorganic Fluorides: Synthesis, Characterization and Properties of Nanostructured Solids* (Chippenham, UK: Wiley)
[2] Nizamutdinov A S et al 2014 Opt. Spectrosc. **116** 732
[3] Csutak S 2008 *Method and Apparatus to Determine Characteristics of an Oil-Based Mud Downhole* US Patent 20080245960 A1
[4] Anderson D F and Sparrow R W 1991 *Divalent fluoride doped cerium fluoride scintillator* US Patent 5039858
[5] Bertaina S et al 2007 Nature Nanotechnology **2** 39
[6] Baibekov E I et al 2017 Phys. Rev. B **95** 064427
[7] Baibekov E I et al 2014 Phys. Rev. B **90**.174402
[8] Aminov L K et al 2013 Phys. Solid State **55** 1343
[9] Aminov L K et al 2006 Appl. Magn. Reson. **30** 175
[10] Ranon U and Hyde J S 1966 Phys. Rev. **141**, 259
[11] Aminov L K et al 2010 Phys. Solid State **52** 2070
[12] Biktagirov T et al 2017 Energy Fuels **31** 1243
[13] Gracheva I N et al 2016 Magnetic Resonance in Solids Electronic Journal **18** 16102
[14] Takui T, Berliner L and Hanson G 2016 *Electron spin resonance (ESR) based quantum computing*. (New York: Springer)

IOP Conf. Series: Earth and Environmental Science **155** (2018) 012004 doi:10.1088/1755-1315/155/1/012004

Study of the oxidized and non- oxidized bitumen modified with additive «Adgezolin» by using electron paramagnetic resonance

I Mukhamatdinov*, M Gafurov, A Kemalov, A Rodionov, G Mamin, and P Fakhretdinov

Kazan Federal University, 18 Kremlevskaya Str., Kazan, Russia
*E-mail: IIMuhamatdinov@gmail.com

Abstract. Cationic surfactant (adhesion additive) «Adgezolin» has been developed. It is shown that introduction of «Adgezolin» into the oxidized bitumen increases the relative amount of asphaltenes and monocyclearomatic hydrocarbons. By means of electron paramagnetic resonance (EPR) it is demonstrated that the introduction of additive «Adgezolin» increases the number of paramagnetic «free» carbon radicals (FR) in the oxidized bitumen and decreases that in the unoxidized species. In both types of bitumen shift from the Lorentzian to Gaussian EPR lineshape of FR is obtained that could be connected with as an increase of the samples homogeneity. It is supposed that while in the oxygenated bitumens introduction of additives leads to the disaggregation of asphaltene-resins compounds, in the unoxidized samples the balance is shifted towards formation of di-radicals.

1. Introduction

The strength and durability of road surface depends largely on the quality of bituminous binder, which allows associating different mineral materials and formation of lasting wearproof and weatherproof monolith (asphalt) [1]. In the Russian Federation oil bitumens and mineral fillers used in the preparation of asphalt mixtures don't have a proper adhesion. This is one of the main causes of premature destruction. Ability to maintain properties during prolonged exposure to high and low temperatures is an important feature for bitumen. To advance bitumen adhesion to the surface of mineral materials asphalt mixtures production technology uses additives (surfactants) [2].

The introduction of additives has a significant influence on the structure of oil dispersed systems (ODS) including a change in composition of oil and structure of the complex structural unit (CSU) composed of asphaltenes and solvate shell mostly composed of resins. Solvate shell includes internal and peripheral parts, consisting of alcohol-benzene and benzene pitches respectively [3, 4]. In this report an influence of the nitrogen-containing surfactant «Adgezolin» adding into the oxidized bitumen BND 60/90 produced by OJSC «TAIF-NK» (denoted hereafter as TAIF, Russia) and unoxidized bitumen BNN 80/120 produced by the oil-and-gas production department of «Elhovneft» (denoted hereafter as Elhov, Russia) are studied.

It is known that one gram of ODS contains up to 10^{20} paramagnetic centers (PC) [7-10]. These include organic free radicals (FR) in the structure of a condensed polyaromatic core of molecules of high-molecular ODs components as well as paramagnetic porphyrin complexes (VP). FR (as well as VP) in ODS are mainly concentrated in asphaltenes and resins and arise due to the delocalized π electrons of the aromatic rings and stable organic radicals of the side chains. The content of the high-molecular ODS components could reach the values of up to 73 wt. % in natural asphalts and bitumen

[8]. Assuming that asphaltenes and resins have molecular weight of about 1000 amu [8], they could contain up to one unpaired electron per molecule. One of the most informative methods to study paramagnetic centers is electron paramagnetic resonance (EPR) (see [8-11] and references therein).

It is assumed that the dispersed colloidal particle stability in ODS petroleum systems is tightly connected with the concentration of PC, first of all of FR [7, 8, 10, 12]. There was many attempts to track the correlations between the EPR spectroscopic parameters and ODS status (content, changes with the oxidation, modification, ageing, temperature, etc.). In this work an attempt to track and connect the obtained changes in EPR spectra with the addition of the produced additive «Adgezolin» is done.

2. Materials and Method

To ensure proper bitumen adhesion to mineral materials, mineral additive are used, in which cationic surface-active agents (surfactants) are the most frequently applied. Nowadays imidazolines and their derivatives [13,14] are the most widely used as additives for bitumen. Polyethylenepolyamines were exploited for the synthesis of «Adgezolin» [5].

Composition of bitumen was evaluated by using a technique of VNII NP [15] based on liquid adsorption chromatography. Maltens could be divided into paraffin-naphthenic (PN-), monocycloaromatic (MCA), bicycloaromatic (BCA), polycycloaromatic (PCA) hydrocarbons (HC) while resins onto the the benzol (BR), alcohol-benzene (ABR) resins, or resin-1 and resin -2, respectively.

EPR spectra were obtained by using ESP 300 spectrometer (Bruker, Germany) operating at 9.4 9.9 GHz (X band) in continuous wave mode with modulation frequency of 100 kHz and amplitude of 0.1 mT. Microwave power level was about 20 μW to avoid saturation effects and to define the EPR parameters correctly. PC concentration was estimated at room temperature in the double cavity ER4105DR by comparing the integrated intensities (I) of the spectra of a test sample and a reference one (Mn(II) in MgO). Heating of the samples in the range of (220-313) K was done in a stream of nitrogen blown through the quartz insert equipped with the VT thermostatic system (Oxford Instruments) with (8 15)°C increment, heating rate of 2°C/min and temperature setting accuracy of not less than ±2°C. To exclude the treatment effect we have measured the PC concentrations at room temperatures before and after heating.

Simulations of the EPR spectra were done by using the functions of the EasySpin module for Matlab software [Stoll et al. 2006] in which the linewidth (Γ) for Gaussian (G) and Lorentzian (L) absorption lineshapes are connected with the full width at half maximum (FWHM) via $\Gamma_G = (2\ln 2)^{-1/2}\cdot$FWHM and $\Gamma_L = (3)^{-1/2}\cdot$FWHM.

3. Results and Discussion

Paraffin-naphthenic hydrocarbons are considered to be undesirable because they dramatically reduce the structural-mechanical strength of bitumen [16]. As it follows from the data presented in Table 1, after adding additives one observes a reduction of PN-hydrocarbons in the TAIF bitumen. Because the mixing of «Adgezolin» with bitumen occurs when heated, this decrease possibly connected with the dehydration of naphthenic rings and formation of alkylaromatic compounds. It results in more lyophilic behavior of asphaltenes [16] which are insoluble in paraffin-naphthenic hydrocarbons, but solvate and swell in aromatic hydrocarbons. Therefore, the asphaltenes can form aggregates. The content of the MCA-hydrocarbons increases almost in 2 times from 6.88% for the original bitumen up to 12.65% for bitumen modified with 1 wt. % of «Adgezolin». Transformation of this faction probably occurs at the expense of dehydrogenation of naphthenic rings in naphtene-aromatic compounds, with the formation of resins and asphaltenes. Thus, the additive adding leads to the increase of both asphaltenes and MCA-hydrocarbons from PN-hydrocarbons:

$$\text{MCA-hydrocarbons} \leftarrow \text{-PN-hydrocarbons} \rightarrow \text{asphaltenes}. \quad (1)$$

Formation of asphaltenes from PN-hydrocarbons may occur through the resin-1 and resin-2:

PN hydrocarbons → resin-1 → resin-2 → asphaltenes. (2)

There is also a possibility avoid the resin formation stage:

PCA HC → MCA HC ← PN HC →resin-1→resin-2 → asphaltenes. (3)

Table 1. Component part of the original and modified TAIF bitumen.

Components, % mass.	Bitumen samples				
	Source	Content of Adgezolin,% of the mass.			
		0.6	0.8	1.0	1.2
P-N HC	17.69	16.17	14.25	12.33	14.20
MCA HC	6.88	8.74	9.77	12.65	12.07
BCA HC	8.26	7.98	7.66	6.85	7.92
PCA HC					
Σ Saturates+aromatics	32.83	32.89	31.68	31.83	34.19
Resin-1	22.82	23.16	21.68	21.81	20.60
Resin-2	17.44	16.31	17.39	18.84	19.32
Σ resins (R)	40.26	39.47	39.07	40.65	39.92
Asphaltenes (A)	21.98	25.36	24.87	25.67	25.89

The results for the unoxidized bitumen Elhov gained from the heavy superviscous oil of Ashal'chinskoe' basin are shown in table 2. With adding additives up to 0.8 % an increase in the amount of saturates + aromatics occurs from 25.90 % for the original bitumen up to 33.50 %. As the number of additives in the bitumen further increases up to 1.2 wt. %, the amount of saturates + aromatics reduces to 29.43 % by reducing the concentration of resin-2. This change is due to the breaking down of resins-2. Asphaltenes may behave in a similar way.

Comparison of the components with the «Adgezolin» adding demonstrates an advantageous transformation of resin-2:

Saturates+aromatics ← resins-2 → asphaltenes. (4)

It is possible that the formation of oils (saturates+aromatics) from resins-2 occurs through the resin-1:

Saturates+aromatics ← resins-1 ← resins-2 → asphaltenes. (5)

Thus, studies in this part of the work demonstrate the influence of the additive «Adgezolin» on the properties and composition of the modified bitumens.

Table 2. Component part of the original and modified Elhov bitumen.

Components,% of the mass.	Bitumen samples				
	Source	Content of Adgezolin,% of the mass.			
		0.6	0.8	1.0	1.2
Saturates+aromatics	25.90	28.57	33.50	29.67	29.43
Resin-1	31.71	36.07	34.47	33.74	36.63
Resin-2	29.09	17.13	15.57	19.21	17.72
Σ resins (R)	60.80	54.20	50.04	52.95	54.35
Asphaltenes (A)	13.31	17.23	16.44	17.38	16.21

EPR patterns in all the studied samples are very similar to each other (Figure 1) and originate from the paramagnetic vanadyl complex VO^{2+} in porphyrins (electronic spin $S = 1/2$, nuclear spin $I = 7/2$; therefore 16 «lines» for the parallel and the perpendicular complex orientations, see [9,17] for details and from FR. Within the accuracy of our experiments, no changes in the spectroscopic parameters of VO^{2+} complexes was obtained with «Adgezolin» addition and temperature dependencies of the integral intensities for VO^{2+} in the investigated samples obeyed Curie law.

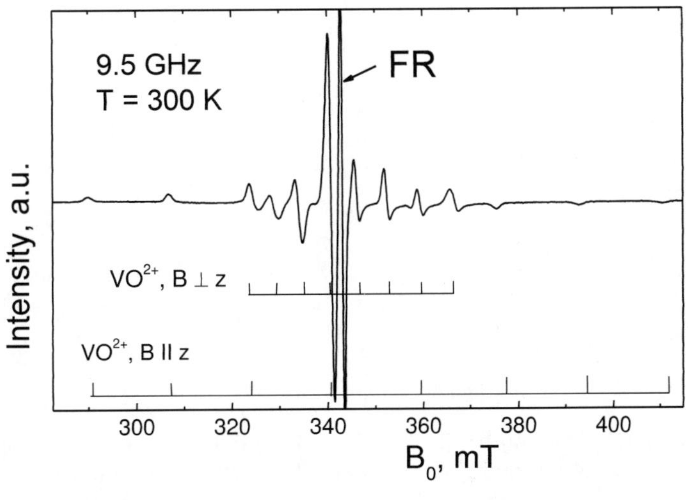

Figure 1. Typical EPR spectrum (TAIF + 0.8 % Adgezolin). Arrow indicates the position of FR; the hyperfine components $A_{\|}$ and $A_{\perp}$ of VO^{2+} corresponding to the parallel and perpendicular orientations in the external magnetic field are shown. The receiver gain is tuned to present all VO^{2+} lines so the FR signal is out of the analog-digital converter range

The EPR spectrum of FR is a single line with the linewidth of $\Delta H_{1/2}$ = (0.54-0.64) mT situated at g-factor 2.003 ± 0.001. FR lineshape is a convolution of Gaussian and Lorentzian in a ratio of about G:L ≈ 34:66 for the untreated samples at RT up to 43:57 for the 1% of (Fig. 2, Table 3). No connection between the oxidation state and additive addition with the line position, width and shape was found. The shift of the lineshape towards the Gaussian form can be regarded as the more homogeneous distribution of «free» radicals in the samples with «Adgezolin» [10].

Table 3. Parameters of approximation for FR line.

Sample	FR concentration, 10^{18} spin/g	VO^{2+} concentration, 10^{18} spin/g	Γ_G, mT	Γ_L, mT	$I_{Gauss}:I_{Lorentz}$ (% : %)
TAIF	1.65(10)	3.6 (5)	0.72	0.76	34:66
TAIF+0.8%	1.85 (10)	3.1 (6)	0.72	0.75	39:61
TAIF+1%	1.95 (15)	3.7 (5)	0.74	0.73	43:57
Elhov	1.50(15)	8.7(5)	0.84	0.82	34:66
Elhov+0.8%	1.10(10)	8.9 (5)	0.84	0.80	42:58
Elhov+1%	1.0(1)	9.7 (5)	0.83	0.81	39: 61

IOP Conf. Series: Earth and Environmental Science **155** (2018) 012004 doi:10.1088/1755-1315/155/1/012004

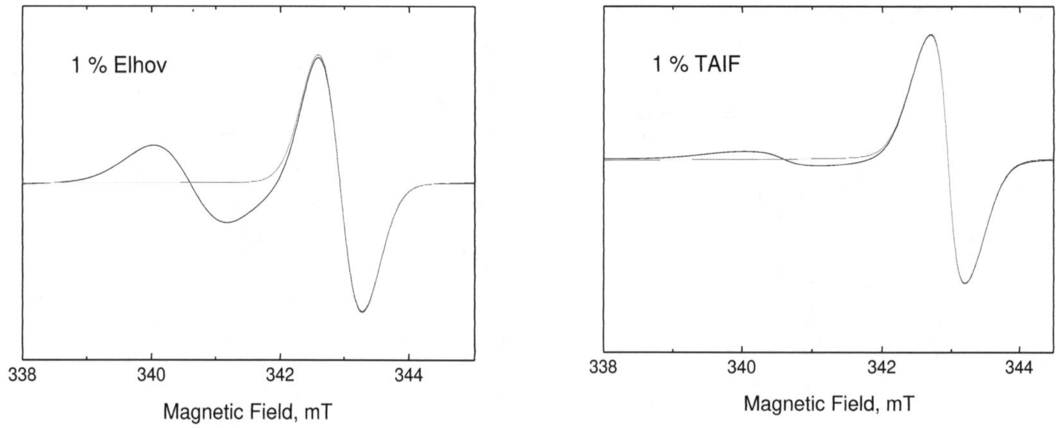

Figure 2. Examples of approximations of FR spectra for 1% Elhov (left panel) and 1% TAIF (right panel). Parts of spectra presenting the most intensive VO^{2+} component and FR line are shown. Details of approximations and their parameters are given in the text

To visualize the *I*(T) dependencies without Curie contributions, the abscissa and ordinate axis are chosen to be 1000/T and (*I*·T), correspondingly. In the case of pure Curie contribution the (*I*·T) would be parallel to the abscissa [8, 10].

As seen from Figure 3, in the temperature range of interest of (220-313) K the EPR intensities for FR practically obey Curie law and can serve as a good number for evaluating FR concentration. Furthermore, because the linewidthes of VO^{2+} and FR only slightly deviates from their values at RT, the peak-to-peak amplitudes (App) of the corresponding EPR lines can be used for this purpose in the same way. That gives a potential opportunity to use EPR as a rapid test for qualitative and quantitative analysis of the surfactant introduction.

From Table 3 it follows that the «Adgezolin» introduction is connected with the significant decrease of the FR numbers in the unoxidized bitumen (Elhov) and increase of those in the oxidized bitumen (TAIF) so that the EPR intensities can differ up to 2 times from sample to sample. This difference, probably, is determined by the nature of asphaltenes and resins. Indeed, it is known that asphaltenes and resins extracted from the bitumens after the high-temperature processing differ from those derived from the natural bitumen. The latter ones are characterized by the lower molecular masses, lower H / C ratio, etc [18]. Increase of amount of FR in the oxidized bitumen, most likely, indicates that the additive «Adgezolin» promotes the dispersion of polar compounds (asphaltenes and resins) resulting in a larger number of the polar particles. In the unoxidized samples «Adgezolin» could also act as a «bridge» for formation of di-radicals which are non-paramagnetic (EPR silent) at least in the investigated temperature range [8, 10].

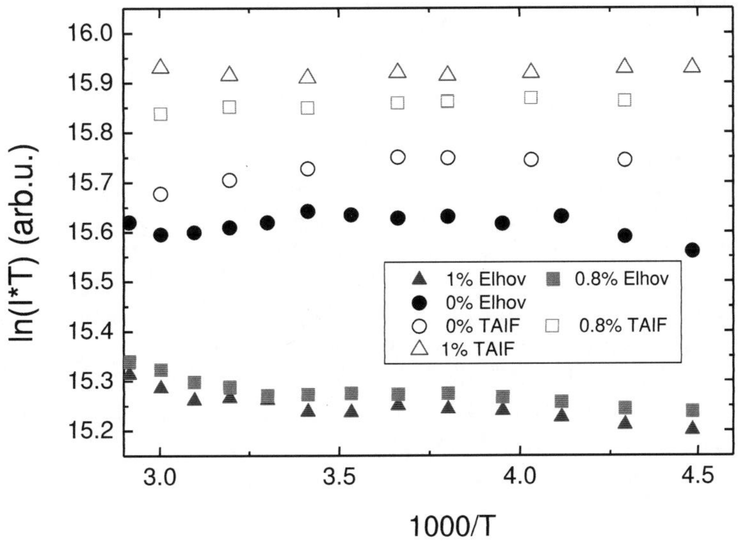

Figure 3. Temperature dependencies of the integral intensities for FR EPR lines for TAIF (open symbols) and Elhov (solid symbols) samples with different content of the adhesion additive «Adgezolin». Horizontal axis is given in a reciprocal temperature (1000/T), the vertical axis is the natural logarithm of product (I·T) to exclude the Curie contribution in determination of FR amount.

4. Conclusion

The establishment of a correlation between the malacometric parameters and a set of parameters determined by using EPR seems to be can help to unravel the mechanism(s) of the modifying action of the «Adgezolin» additive. Namely, the optimal dosage, the way of the additive introduction into the asphalts of various origins, etc., can be derived from these kinds of measurements. This can be used to develop an express method for the rapid control of content and for the qualitative analysis of the modified bitumen.

Acknowledgments

The work was funded by the subsidy of the Russian Government to support the Program of competitive Growth of Kazan Federal University among World's Leading Academic Centers.

References

[1] Pasandín A R and Pérez I 2015 *International Journal of Adhesion and Adhesives* **58** 53
[2] Cui S, Blackman B R K, Kinloch A J, and Taylor A C 2014 *International Journal of Adhesion and Adhesives* **54** 100
[3] Syunjaev Z I, Sjunjaev R Z, and Safieva R Z 1990 *The oil disperse systems* p. 226 (in Russian)
[4] Glagoleva O F 2012 *Refining and petrochemicals* **6** 16 (in Russian)
[5] Mukhamatdinov I I, Kemalov A F, and Fakhretdinov P S 2017 *Chemistry and Technology of Fuels and Oils* **53** 683
[6] Mukhamatdinov I I, Fakhretdinov P S, and Kemalov A F 2017 *17th International Multidisciplinary Scientific GeoConference SGEM-2017* **17** 575
[7] Syunyaev Z I 1980 *Chemistry and Technology of Fuels and Oils* **16** 484

[8] Yen T F and Chilingarian G V 2000 *Asphaltenes and asphalts.* Vol. 2 *(Developments in petroleum science)* 40B (Amsterdam: Elsiever Science) p 622
[9] Biktagirov T, Gafurov M, Volodin M, Mamin G, Rodionov A, Izotov V, Vakhin A, Isakov D, and Orlinskii S B 2014 *Energy Fuels* **28** 6683
[10] Dolomatov M U, Rodionov A A, Gafurov M R, Petrov A V, Biktagirov T B, Bakhtizin R Z, Makarchikov S O, Khairudinov I R, and Orlinskii S B 2016. *Magnetic Resonance in Solids*, **18** 16101
[11] Gafurov M, Volodin M, Rodionov A, Sorokina A, Dolomatov M, Petrov A, Vakhin A, Mamin G and Orlinskii S 2018 *Journal of Petroleum Science and Engineering* **166** 363
[12] Pokonova Y V 1980 *Chemistry of high-molecular compounds of the oil* p 172(in Russian)
[13] Schramm L L, Stasiuk E N, and Mazangoni D C 2003 Annu. Rep. Prog. Chem., Sect. C: Phys. Chem., **99** 3
[14] Rubingh D N, and Holland P M 1990 *Cationic surfactants: physical chemistry* p 544
[15] Gong R B 1973 *Oil bitumen* p 432 (in Russian)
[16] Rozental' D A, Syroezhko A M, and Ralis R V 2007 *Petroleum Chemistry* **47** 299
[17] Mamin G V, Gafurov M R, Yusupov R V, Gracheva I N, Ganeeva Yu M, Yusupova T N, and Orlinskii S B 2016 *Energy&Fuels* **30** 6942
[18] Lesueur D 2009 *Adv Colloid Interface Sci.* **145** 42

IOP Conf. Series: Earth and Environmental Science **155** (2018) 012005 doi:10.1088/1755-1315/155/1/012005

W-band EPR of vanadyl complexes aggregates on the surface of Al_2O_3

G. Mamin*, M. Gafurov, A. Galukhin, I. Gracheva, F. Murzakhanov, A. Rodionov, and S.Orlinskii

Kazan Federal University, 18 Kremlevskaya Str., Kazan, Russia
E-mail: george.mamin@kpfu.ru

Abstract. Structural characterization of metalloporphyrins, asphaltenes and their aggregates in complex systems such as native hydrocarbons is in the focus of scientific and industrial interests since many years. We present W-band (95 GHz) electron paramagnetic resonance (EPR) study in the magnetic field of about 3.4 T and temperature of 100 K for Karmalinskoe oil, asphaltens and asphaltenes deposited on the surface of Al_2O_3. Features of the obtained spectra are described. Shift to the higher frequencies allows to separate spectrally the contributions from paramagnetic complexes of different origin and define the EPR parameters more accurately comparing to the conventional X-band (9 GHz). Changes of the EPR parameters are tracked. We suggest that the proposed approach can be used for the investigation of structure of vanadyl complexes aggregates in crude oil and their fractions.

1. Introduction

Paramagnetic properties and electron paramagnetic resonance (EPR) spectra in complex hydrocarbon systems (like oil rocks, crude oil, bitumen, their fractions, etc. which we will denote in the text as PDS – petroleum disperse system) are mainly defined by the presence of complexes of *d*-metals (first of all V, Ni, Fe) and stable "free" radicals (FR). It is usually assumed that FR are mainly concentrated in the heaviest PDS fraction - asphaltenes and arise due to the delocalized π-electrons of the aromatic rings and stable organic radicals of the side chains. One has to note that the exact location and structure of FR in PDS are still undefined and the researchers are conventionally referred to the paramagnetic centers detected in coals [1, 2]. Therefore, it would not be considered in details in this work.

Another important type of paramagnetic centers in PDS which is detected in asphaltenes and resins fractions are vanadyl based porphiryns. Metalloporphiryns in heavy oils were first discovered in 1930s by Treibs [3, 4], who postulated the biological origin of these compounds. Vanadyl (VO^{2+}) porphyrin is one of the most abundant forms of metalloporphyrin, and is of particular significance as a biomarker in the context of oil formation research [5]. PDS porphyrins exist in homologous manifolds of several structural classes and manifest great structural diversity [6, 7]. The molecule can contain different types of substitutions, including alkyl, cycloalkane, and aromatic groups and possesses near planar anisotropic structure [8-10].

The possibility of using the paramagnetic VO^{2+} complexes as a probe for studying complex disordered systems such as PDS is determined by their resolved anisotropic structure in the electron paramagetic resonance (EPR) spectra [9-11] which can be also further spectrally separated from FR EPR by shifting to high frequencies/ high magnetic fields [12, 13]. Nevertheless, there were no many papers on this issue in the near past [14, 15]. Commercial availability of the pulsed high field EPR spectrometers open new ways for the PDS analysis, investigation of the asphaltenes' structure, processes of their treatment, etc. using intrinsic paramagnetic centers as sensitive probes [9, 10, 12, 16-18].

One of the advantages of EPR is that species can be investigated without any additional treatment, in native environment. In paper [10] we have demonstrated that W-band EPR and ENDOR of VO^{2+} containing complexes are very similar for two very different crude oils. From other side, in paper [11] we have shown that the temperature dependencies of X-band (9 GHz) VO^{2+} EPR spectra at T > 400 K are different for the same oil samples. It can serve as a sign that VO^{2+} complexes are the part of stable asphaltene structure/structures which are universal at least for some types of PDS and VO^{2+} is a sensitive native paramagnetic centre to probe the PDS structure.

In the past [19, 20] the deposition of monolayers or double layers of model ("pure") VO^{2+} complexes like V_2O_5, for example, was widely used to investigate and influence the active sites of different catalytic supports, including alumina (Al_2O_3) one. In this research we use the inverse approach - an attempt to unravel some features of the VO^{2+} complexes aggregates in PDS by using asphaltenes deposition on the surface of Al_2O_3. Our own experience of characterization of Al_2O_3 bulk and surfaces with EPR/ENDOR along with the corresponding literature reviews are gathered in ref. [21-23].

The effect of surrounding matrices, temperature, catalytic treatments, etc. on model VO complexes and those in maltenes, resins, asphaltenes in EPR is conventionally accounted through the small deviations of components of g-factors between each other and from the value of g-factor of free electron (see detailed explanation in Section 3) [19, 20, 24]. Obviously, for the high frequency experiments such shift and differences should be more pronounced. But as it was already mentioned, no experimental evidence for that was provided.

2. Materials and Methods

Crude oil was extracted from the oil core of the Karmalinskoe oilfield (Volga-Ural basin, Republic of Tatarstan, Russia). Compositional analysis of the oil sample (SARA analysis) was conducted according to the common technique [25] which includes asphaltenes precipitation from heptane and further chromatographic separation of the non-asphaltic oil components through column filled with alumina. SARA (saturated, aromatic, resins) analysis data of the studied crude oil are presented in Table 1. Organic solvents, such as, toluene, heptane, dichloromethane, methanol (purity of all solvents was no less than 99.5%) were purchased from Component-Reaktiv and used without additional purification. Aluminum oxide (particle size 0.05-0.15 mm, neutral) were purchased from Sigma Aldrich.

Table 1. Results of SARA analysis for the crude oil sample.

SARA analysis, %			
Saturated	Aromatic	Resins	Asphaltenes
19.4	45.8	29.0	8.4

To prepare sample with the adsorbed asphaltenes (As), 1.000 g of Al_2O_3 was added to dichloromethane solutions containing 2×10^{-3} g×ml^{-1} of As in a round bottom flasks. Obtained mixtures were shacked for 30 min, and then Al_2O_3 with adsorbed asphaltenes was filtered off and dried in vacuum evaporator under reduced pressure (20 mbar, 40 °C). The resulting sample we will denote as Q30A. The calculations show that the deposited As amount corresponds to mono or double layer on the aluina surface.

W-band EPR spectra at T = 100 K were registered with Bruker ElexSys 680 EPR spectrometer by using standard electron-spin-echo (ESE) $\pi/2 - \tau - \pi$ with the $\pi/2$ pulse duration of 32 ns, time delay $\tau = 240$ ns pulse sequence. The ability to detect ESE gives an opportunity to use more advanced pulsed techniques in the near future. X-band concentration measurements are done by using Bruker

ESP-300 with double cavity ER4105DR by comparing the integrated intensities of the spectra of a test sample and a reference sample (powder of DPPH and Mn^{2+} in MgO). Spectra simulation are done in Easyspin package for MatLab [26].

3. Results and Discussion.

Initial Al_2O_3 samples are EPR silent within the sensitivity limit of our instruments. Comparison of the W-band ESE spectra for crude oil, As powder and deposited Q30A sample are shown in Figure 1. The EPR spectra contain two contributions: the first one is related to VO^{2+} complex ($V^{4+}+O^{2-}\rightarrow VO^{2+}$) whereas the second (single line) can be attributed to "free" radical(s). As it was already mentioned, shift to the W-band allows In this work we are concentrating on the tracing of EPR of VO^{2+} complex in the W-band. EPR spectrum of VO^{2+} at X-band is described in details in [11].

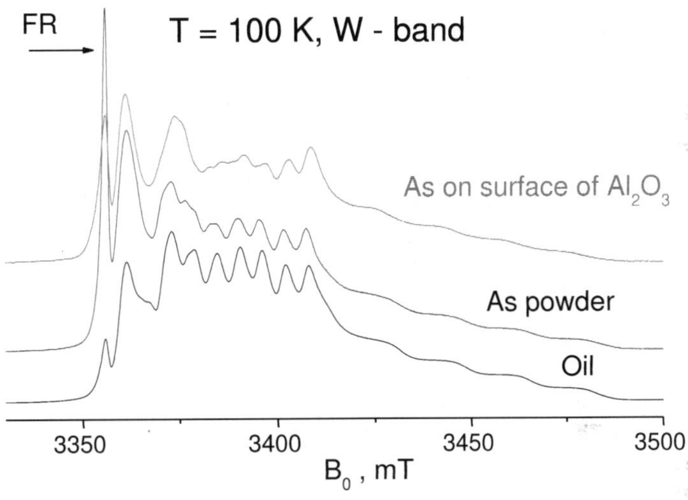

Figure 1. The W-band ESE spectra at T = 100 K of initial oil sample (the lower curve), As sample (in the middle) and Q30A (top line). Arrow indicates the line due to FR. Other parts of spectra are due to VO^{2+} paramagnetic complexes.

EPR spectra of Oil and As are very similar. It allows suggesting that As structure remains the same as in the initial oil sample after the extraction process. Alteration of EPR pattern for the deposited Q30A sample is glaringly obvious but it is hard to describe numerically. To follow the changes for the deposited sample, we have to characterize the EPR spectra in details (see Figure 2).

The largest interaction in the high-field experiments is the Zeeman one between the magnetic moment of the unpaired electronic spin S and B_0. The VO^{2+} paramagnetic complex is characterized by $S = ½$ and, therefore, by two projections of $m_S = \pm 1/2$ onto the direction of B_0. The electron Zeeman interaction is described by the parameter called g-factor with the typical values close to value 2.00. Another interaction (common applicable to hydrocarbon systems) is the spin-nuclear one. As a nucleus could also have a magnetic moment, the interaction of the unpaired electron with the nucleus splits the electron energy levels, generating a structure called hyperfine structure in the EPR spectrum – each energy level splits into a closely spaced group of $(2I+1)$ sublevels (here I is the nuclear spin quantum number), described by the hyperfine parameter A. For ^{51}V nuclear spin $I = 7/2$.

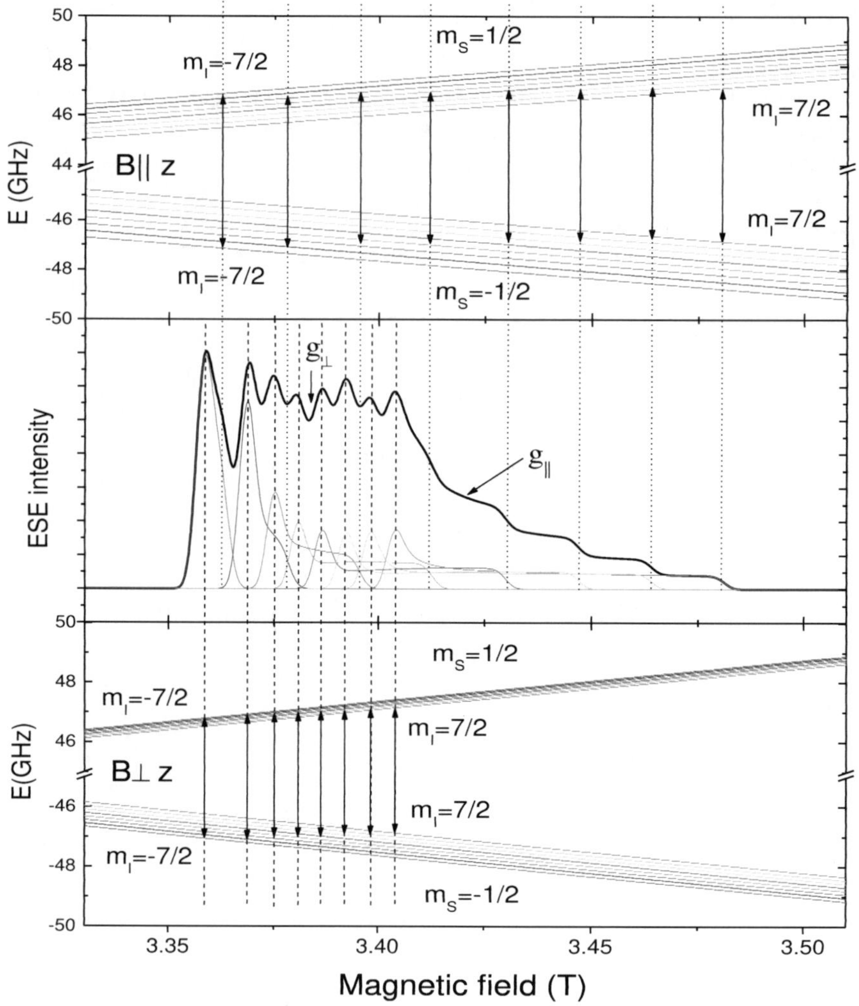

Figure 2. The energy levels and the corresponding absorption EPR spectrum for VO^{2+} complex calculated for the microwave frequency $\nu = 94$ GHz, $g_{||} = 1.963$, $g_{\perp} = 1.985$, $A_{//} = 470$ MHz, $A_{\perp} = 150$ MHz. Particular contributions from every EPR transition are color marked. Calculations are done in EasySpin package for Matlab [26].

Taking into account that VO^{2+} has a near planar symmetry, the *g*- and *A*-parameters should be regarded as anisotropic parameters having axial symmetry, i.e, $(g, A)_{||}$ and $(g, A)_{\perp}$. Figure 2 presents the calculated electron-nuclear energy levels of VO^{2+} paramagnetic complex, EPR transitions and the corresponding EPR absorption line with the values of $(g, A)_{||}$ (i.e., perpendicular to VO^{2+} plane, *z* axis) and $(g, A)_{\perp}$ (i.e., in the VO^{2+} plane or *xy* plane) taken from [11] which will be appeared close to the obtained in the experiment (Table 2). As for the most VO^{2+} related investigations, we suppose that the direction of the main axes coincide for *g*- and *A*- tensors. The powder EPR spectrum of VO^{2+} represents the 2×8 hyperfine patterns (the projection of *I* is allowed to take 8 values: of m_I: ±7/2; ±5/2;

±3/2; ±1/2) for the parallel and the perpendicular complex orientations. Only transitions are allowed with $\Delta m_S = \pm 1$ and $\Delta m_I = 0$.

As it can be seen from the comparison of Figures 1 and 2, experimental spectra only with some extend of accuracy can be described by the spin-Hamiltonian of the axial symmetry for the single VO^{2+} molecule. The extracted from the experimental spectra values of *g*- and *A* – components are presented in Table 2. Due to the higher frequency (the higher spectral resolution) one can define the components of g-factors with much better accuracy than in the conventional X-band. From Table 2 it follows that the transition from the bulk sample to the deposited one is accompanied by some increase of $g_{||}$ and decrease of $A_{||}$.

Table 2. Values of of g- and *A*- components for VO^{2+} paramagnetic centers at room temperature as measured in the W-band.

	$g_{\|\|}$	$g_{\perp}$	$B = \Delta g_{\|\|}/\Delta g_{\perp}$	$A_{\|\|}$, MHz	$A_{\perp}$, MHz
Oil	1.9610(8)	1.9825(5)	2.1	480(10)	155(5)
As (powder)	1.9610(8)	1.9825(5)	2.1	475(15)	155(5)
Q30A	1.9630(10)	1.9820(5)	1.9	460(15)	155(5)

The first one is in the accordance with ref. [20] in which double and monolayers of V_2O_5 on different surfaces were considered. Usually, for accounting of g-factor changes numerically, the B parameter has been used as a sensitive indicator of structural distortions to clarify a change in the V=O bond length which can be derived as follows:

$$B = \Delta g_{||}/\Delta g_{\perp}, \quad (1)$$

where $\Delta g_{||} = g_{||} - g_e$, $\Delta g_{\perp} = g_{\perp} - g_e$ and g_e is the free electron g value of 2.0023.

Definitely, the B parameter for the deposited sample is slightly, but larger than that in the initial asphaltene. For simple vanadyl complexes it is ascribed to a shorter V=O bond and/or a longer distance to the nitrogen ligands, i.e., the higher electron density in the axial bond and more electron delocalization to the V orbital [24]. But, in our opinion, it should be accompanied with the decrease of *A*-components especially in the plane of V=O bond, i.e. of $A_{\perp}$ which was not observed; neither in the referenced paper [24] nor in our experiments. Instead that we have obtained the reduction of the out-of-plane parameter $A_{||}$ (see Table 2). It confirms additionally that changes of the EPR parameters in PDS are mainly due to the disaggregation of the complex vanadyl aggregates rather than due to the structural changes of single VO containing molecule. We hope to address this issue in details by using the more elaborated EPR techniques like electron-nuclear double resonance (ENDOR) [27].

4. Conclusion

Until recently a role of metalloporphyrins in the aggregation of the PDS high-molecular weight phases was discounted or even denied. The latter experimental and theoretical studies show that vanadyls can be of crucial importance for formation of the nano-, micro- and macro aggregates [28, 29]. We hope, that the proposed approach for studying the vanadyl aggregates could help to unravel also the nature of asphaltene aggregates

Acknowledgments
The work was done in the line with the Program of competitive Growth of Kazan Federal University among World's Leading Academic Centers and supported by RFBR grant # 18-42-160016 p_a.

References

[1] Yen T F and Chilingarian G V 2000 *Asphaltenes and asphalts.* Vol. 2 *(Developments in petroleum science)* 40B (Amsterdam: Elsiever Science) p 622
[2] Yen T F and Chilingarian G V 1994 *Asphaltenes and Asphalts.* Vol. 1 (*Developments in Petroleum Science*) 40A (Amsterdam: Elsiever Science) p 459
[3] Treibs A 1934 *Ann. Chem.* **510** 42
[4] Treibs A 1936 *Angew. Chem.* **49** 682
[5] Baker E and Louda J 1986 in *Biological Markers in the sedimentary record* ed. JohnR B (Amsterdam: Elsiever) 125-225
[6] Zhao X, Liu Y, Xu C et al. 2013 *Energy Fuels* **27** 2874
[7] Zhao X, Shi Q, Gray M,R and Xu C 2014 *Scientific Reports* **4** 5373
[8] Dechaine G P and Gray M R 2010 *Energy Fuels* **24** 2795
[9] Gracheva I N, Gafurov M R, Mamin G V, Biktagirov T B, Rodionov A A, Galukhin A V and Orlinskii S B 2016 *Magn. Reson. Solids* **18** 16102
[10] Biktagirov T, Gafurov M, Mamin G, Gracheva I, Galukhin A and Orlinskii S 2017 *Energy Fuels* **31** 1243
[11] Biktagirov T B, Gafurov M R, Volodin M A, Mamin G V, Rodionov A A, Izotov V V, Vakhin A, Isakov D and Orlinskii S B 2014 *Energy Fuels* **28** 6683
[12] Mamin G V, Gafurov M R, Yusupov R V, Gracheva I N, Ganeeva Y M, Yusupova T N and Orlinskii S B 2016 *Energy Fuels* **30** 6942
[13] AlexandrovA S, Archipov R V, Ivanov A A, Gnezdilov O I, Gafurov M R and Skirda V D 2014 *Appl. Magn. Reson.* **45** 1275
[14] Atherton N M, Fairhurst S A and Hewson G J 1987 *Magn. Reson. Chem.* **25** 829
[15] Galtsev V E, Ametov I M and Grinberg O.Ya. 1995 *Fuel* **74** 670
[16] Ramachandran V, van Tol J, McKenna A M, Rodgers R P, Marshall A G and Dalal N S 2015 *Anal. Chem.* **87** 2306
[17] Ben Tayeb K, Delpoux O, Barbier J, Marques J, Verstraete J and Vezin H 2015 *Energy Fuels* **29** 4608
[18] Ben Tayeb K, Delpoux O, Barbier J, Chatron-Michaud P, Dogne M and Vezin H 2017 *Energy Fuels* **31** 3288
[19] Silbernagel B G 1979 *J. Catal.* **56** 315
[20] Sharma V K, Wokaun A and Baiker A 1986 *J. Phys. Chem.* **90** 2715
[21] Gafurov M R, Mukhambetov I N, Yavkin B V, Mamin G V, Lamberov A A and Orlinskii S B 2015 *J. Phys. Chem. C* **119** 27410
[22] Mukhambetov I N, Lamberov A A, Yavkin B V, Gafurov M R, Mamin G V and Orlinskii S B 2014 *J. Phys. Chem. C* **118** 14998
[23] Mamin G V, Orlinskii S B, Rodionov A A and Tagirov M S 2015 *JETP Lett.* **102** 628
[24] Cui Q, Nakabayashi K, Ma X et al. 2017 *RSC Advances* **7** 37908
[25] Galukhin A V, Erokhin A A, Osin Y N, Nurgaliev D K 2015 *Energy Fuels* **29** 4768
[26] Stoll S and Schweiger A 2006 *J. Magn. Reson.* **178** 42
[27] Yavkin B V, Mamin G V, Orlinskii S B et al. 2012 *Phys. Chem. Chem. Phys.* **14** 2246
[28] Trukhan S N, Kazarian S G and Martyanov O N 2017 *Energy Fuels* **31** 387
[29] Silva H S, Sodero A C R, Korb J-P, Alfarra A, Giusti P, Vallverdu G, Bégué D, Baraille I and Bouyssiere B 2017 *Fuel* **188** 374

IOP Conf. Series: Earth and Environmental Science **155** (2018) 012006 doi:10.1088/1755-1315/155/1/012006

Conventional electron paramagnetic resonance of Mn^{2+} in synthetic hydroxyapatite at different concentrations of the doped manganese

F Murzakhanov[1,*], G Mamin[1], A Voloshin[1], E Klimashina[2], V Putlyaev[2], V. Doronin[3], S. Bakhteev[3], R. Yusupov[3], M Gafurov[1], and S Orlinskii[1]

[1]Kazan Federal University, 18 Kremlevskaya Str., Kazan, Russia
[2]Lomonosov Moscow State University, Leninski Gori, Moscow, Russia
[3]Kazan National Research Technological University, 41 K. Marxa Str., Kazan, Russia

E-mail: murzakhanov.fadis@yandex.ru

Abstract. Powders of synthetic hydroxyapatite doped with Mn^{2+} ions in concentrations from 0.05 till 5 wt. % were investigated by conventional electron paramagnetic resonance (EPR). The parameters of the spin-Hamiltonian are derived. Partially resolved hyperfine structure in the magnetic fields corresponding to $g \approx 4.3$ and $g \approx 9.4$ is observed. The narrowing of the central peak with concentration is reported. A possibility to use the linewidth and intensity of the central peak for concentration measurements are discussed. The results could be used for the identification and qualification of Mn^{2+} in oil, mining and ore formations.

1. Introduction

Calcium phosphates (CaP) are the materials of the scientific and application interest due to their abundance in the nature and presence in the living organism. In nature, different calcium phosphate minerals are produced within a wide range of environments by geological (igneous apatite), geochemical and/or geomicrobiological (phosphorite), and biological (biological apatite) processes. Igneous apatite minerals nucleate and crystallize from molten, phosphate-rich rock, forming crystalline fluorapatite ($Ca_5[PO_4]_3F$), chlorapatite ($Ca_5[PO_4]_3Cl$), or hydroxyapatite ($Ca_{10}[PO_4]_6[OH]_2$, HAp or HA) [1]. The experimental results indicate that HAp/MnO_2 composite may be an effective adsorbent for the removal of lead ions from aqueous solutions [2], while the HAp supported manganese could serve as precursor for the oxide catalysts [3].

Manganese is essential impurity element in CaP. Oil containing formation – rock, sands or bitumen very often possess impurity manganese [4]. One of the powerful experimental methods for identification of manganese, especially in 2+ state (Mn^{2+}) in different types of materials (including apatites) and characterization of structure of their complexes is electron paramagnetic resonance (EPR) [4-16]. Despite the large number of the studies performed, many important problems relating to anionic and cationic substitutions in HAp are not sufficiently investigated. Moreover, data of different studies are contradictive. The most contradictive information is associated with the sites of the ions localization in biomineral, synthetic and nanosized samples.

In this work we investigate Mn^{2+} doped synthetic HAp with the nominal concentrations of manganese in the range of 0.05 wt % - 5 wt % by using conventional EPR approaches at X-band range (9 GHz) aimed to show some additional (or not well-known) abilities of the established technique for characterizing CaP:Mn systems.

2. Materials and Methods

Mn^{2+}- containing HAp samples with the chemical formula $Ca_{10-x}Mn_x(PO_4)_6(OH)_2$ (Mn-HAp) were synthesized by the wet precipitation technique. Calcium nitrate monohydrate $Ca(NO_3)_2{\bullet}H_2O$ (235 g) was dissolved in 420 ml of 20% NH_4OH (Solution A) while $(NH_4)_2HPO_4$ (72.2 g) was dissolved in 380 ml of deionized water (Solution B). After the $(NH_4)_2HPO_4$ was completely dissolved, 30 ml of 20% NH_4OH was added. Mn-containing reagent ($MnSO_4{\bullet}H_2O$) was dissolved in 100 ml of deionized water and then added to the A solution. The B solution was then added to the Mn-containing A solution and mixed by a magnetic stirrer for 24 h. After mixing, the precipitate was allowed to age for 48 h. Subsequently, the supernatant was decanted and then washed with 2.5 l of deionized water three times to remove the excess of NH_4OH and NO_3^-. The washing procedure was repeated three times. After washing, the Mn-HAp precipitate was separated from the liquid by filtering through Buhner funnel. A part of the Mn-HAp precipitate was heated at 900°C for 3 h to remove residual NH_4NO_3 and improve sample crystallinity. In this work we investigate four samples with the nominal concentrations of 0.05; 0.1; 1 and 5 wt %.

X-ray diffraction (XRD) analysis (D2 Phaser diffractometer, Bruker) shows that all the samples contain only one phase with the parameters of the unit cell typical for the bulk crystals of the hydroxyapatite. The micromorphology of the powders was examined by scanning and transmission electron microscopy (JEM-2000FX II, JEOL, operated at 200 kV; and FESEM LEO SUPRA 50VP, Carl Zeiss, 5 kV). The chemical content of the powders was defined by EDS (INCA Energy+, Oxford Instruments, attached to LEO SUPRA 50 VP), ICP-MS (ELAN - DRC II, Perkin Elmer) and X-ray fluorescence analysis with total external reflection (S2 Picofox, Bruker).

Continuous wave EPR measurements were done at room temperature using X-band (9 GHz) Varian E-12, Bruker ESP-300, Elexsys 580 and table top LABRADOR (Ekaterinburg, Russia) spectrometers to compare the results obtained at different equipment and test limitations of each of them.

3. Results and Discussion

EPR spectra for the investigated samples are presented in Figure 1. One of the remarkable features of EPR (in addition to the facts that it is a non-destructive method with the high selective sensitivity to Mn^{2+}complexes) that its parameters are sensitive to the interaction of paramagnetic centers with the magnetic moments of the own and neighbouring nuclei (hyperfine and super-hyperfine interaction). Because the nuclear spin I = 5/2 for ^{55}Mn, six lines (2I + 1 = 6) of partially resolved hyperfine (HF) structure of Mn^{2+} with the isotropic HF constant $A_{iso} \approx 90(2)$ G are to obtain at top of the central one corresponding to g = 2.01 (Figure 1 A). To make a transition from the magnetic field values to the frequency ν and to the dimensionless g-factor, a standard EPR equation for the Zeeman interaction with the magnetic field:

$$h\nu = g\beta B, \qquad (1)$$

where h is a Planck constant, β is an electronic Bohr magneton, B is a magnetic field strength is used.

The investigated system could be described by the spin-Hamiltonian:

$$\hat{H} = g\beta \mathbf{BS} + D*[S^2_z - S*(S+1)/3] + E*(S^2_x - S^2_y) + A\mathbf{SI}, \qquad (2)$$

where $S_{x,y,z}$ are the projections of the electronic spin S = 5/2 on the principal direction of D, D and E are the commonly used (axial and orthorhombic, correspondingly) parameters for zero-field interaction (interaction with the crystal field), other parameters are defined above in the text. For simplicity we assume that the main axes of g, A and D tensors coincide. To describe the powder spectra function *pepper* of the EasySpin program was used [17]. It gives g = 2.01, D = 530(40) G,

E = 145(25) G, A_{iso} = 90±2 Гс. It follows that E/D ≈ 0.27, i.e. the system is quite anisotropic like glass [18]. Nevertheless, the g and A tensors could be taken isotropic and XRD analysis (within its accuracy) does not reveal any traces of the amorphous phases.

The presented in Figure 1 result is not trivial. Very often the observation and interpretation of the HF structure (HFS) is a rather complicated task. Even in the perfect crystals the resolution of the HFS spectra, the number of the components of the structure, their relative intensities strongly depend on the concentration of PC, orientation of the applied magnetic field and the EPR frequency [8-10, 19]. That is why, for example, the resolved HFS of Mn^{2+} is rarely observed at X-band in the disordered systems (powders, ceramics, glasses, nanosized samples, inhomogeneous samples of biogenic origin, samples containing non-crystal amorphous phase, etc) and a transition to the higher frequencies are necessary to interpret the EPR spectra correctly (Figure 1, ref. [8-10, 15]).

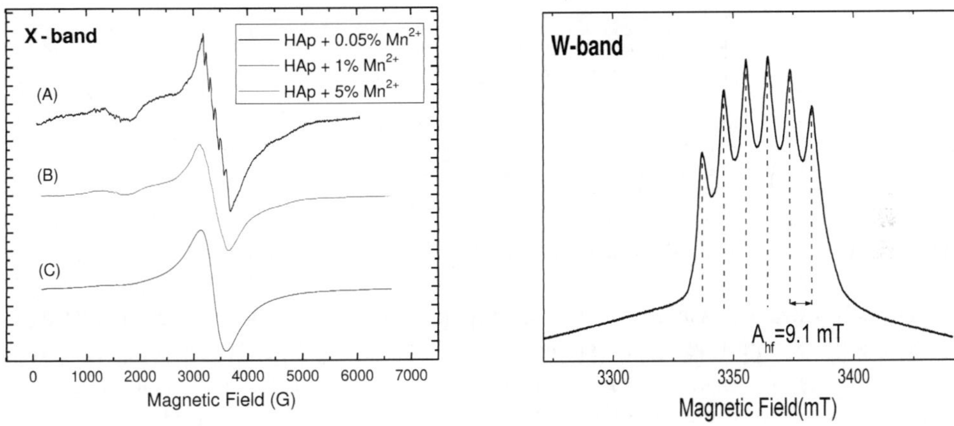

Figure 1. Left panel: comparison of EPR spectra of the manganese doped HAp at 9 GHz and T = 300 K for (A) 0.05%; (B) 1 % and (C) 5 %. Six lines of the partially resolved hyperfine structure due to the I = 5/2 for ^{55}Mn nuclei is to observe for 0.05% Mn^{2+}. The signal amplitudes (vertical axes) are normalized for visibility. Right panel: W-band (95 GHz) field-swept electron spin echo EPR spectrum of 0.1% Mn-HAp detected at room temperature. The components of electron-nuclear hyperfine splitting are marked.

Furthermore, we observed HFS not only on the central line but also in the magnetic fields corresponding to g ≈ 4.3 and g ≈ 9.4 (Figure 2) in samples with the lowest concentration of Mn^{2+}. Such features were rarely reported in the literature (especially for g ≈ 9.4) [20]. Conventionally, the appearance of the EPR signals at low magnetic fields in glasses is ascribed to the large values of D and E parameters, while the possibility to obtain the resolved HF in low magnetic fields is speculated to be due to the two definite positions in the matrix with the distinct but quite specified values of D and E [18]. Following this line of reasoning, our experimental results confirm that manganese at already low concentrations occupies both of the two different calcium positions in HAp [12].

From the application point of view, the obtained feature could help to interpret the results of CaP inspection by EPR. Indeed, the detection of EPR signals at g ≈ 2, g ≈ 4.3 and sometimes at g ≈ 9 in geological and biological samples is conventionally ascribed to the Fe^{3+} ions [11]. But the most abundant stable isotopes of Fe either have I = 0 (^{56}Fe and ^{58}Fe) or I = ½ (^{57}Fe) and could not give six line HF pattern.

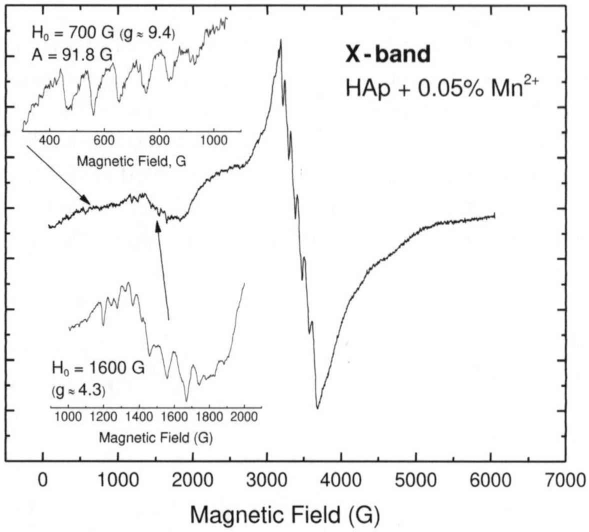

Figure 2. EPR of 0.05 % Mn^{2+} with the amplified parts corresponding to $g \approx 4.3$ and $g \approx 9.4$.

At higher concentrations electron-electron spin-spin interaction leads to the broadening of the EPR lines, HFS disappears, the lines in the vicinities of $g \approx 4.3$ and $g \approx 9.4$ flatten, only central peak which mainly corresponds to the allowed transition (with $\Delta Ms = -1/2 \leftrightarrow + ½$ around $g \approx 2.0$) remains dominate. Analysis of its lineshape shows that for 5 % Mn^{2+} it is well fitted by the only Lorentzian curve while at the lower concentrations small Gaussian contribution appears. The central peak narrows with concentration from 1190 G down to 920 G for the halfwidth on the half peak, $\Delta H_{1/2}$ that could be used for the rough estimation of Mn^{2+} concentration or its change during the release - adsorption processes in catalytic processes, for example. The observed narrowing in paper [8] was connected with the exchange interaction through the oxygen bridges. Exact analytical equation for describing the obtained behavior is still not developed.

Excellent coincidence between intensity of only the central EPR peak (after double integration) with ICP-MS data (Table 1) confirms that this simple way of estimation of Mn^{2+} concentration is applicable at least for the investigated concentration range. This is of particular importance for the specialists who use small - type EPR spectrometers like LABRADOR or Bruker e-scan: they are able to sweep the magnetic field only in the narrow field range (as a rule around $g \approx 2$). As we show even that is enough to estimate the concentration of Mn^{2+} in HAp correctly.

Table 1. Comparison of ICP-MS data with the relative intensity of the central EPR line after the double integration.

Nominal concentration, wt. %	M_{Mn}/M_{Ca} from ICP-MS (relative values)	Relative integral intensity of the central EPR line
0.05 %	1	1
0.1 %	1.86(6)	1.8(1)
1%	18.53(4)	19(1)
5%	99.69(5)	96(3)

4. Conclusion and Perspectives

In this paper we have shown that the possibility of even conventional EPR approaches for investigation and analysis of Mn doped HA are still not fully exploited and many mechanisms of the spin-spin interaction in HA (like exchange interaction) are not described in details. We hope that our paper would encourage the specialists in different scientific and industrial branches to use different EPR approaches in their studies and routine analyses.

Acknowledgements

This work was supported by Russian Foundation for Basic Research under Grant # 18-42-160015 p_a. E.K and V.P. acknowledge partial support from Lomonosov Moscow State University Program of Development.

References

[1] Omelon S, Ariganello M, Bonucci E, Grynpas M and Nanci A 2013 *Calcif. Tissue Int.* **93** 382
[2] Dong L, Zhu Z, Qiu Y and Zhao J 2016 *Front. Environ. Sci. Eng.* **10** 28
[3] Chlala D, Giraudon J M, Nuns N, Lancelot C, Vannier R N, Labaki M and Lamonier J F 2016 *Appl. Catal. B Environ.* **184** 87
[4] Khasanova N M, Gabdrakhmanov D T, Kayukova G P, Mikhaylova F N and Morozov V P 2017 *Magn. Reson. Solids* **19** 17102
[5] Low J 1957 *Phys. Rev.* **105** 793
[6] Gilinskaya L G and Sherbakova M Y 1971 *J. Struct. Chem.* **11** 950
[7] Gilinskaya L G and Sherbakova M Y 1975 *Isomorphous substitution and structural defects in apatites as found by EPR investigation* (Novosibirsk: Publishing House "Nauka") p 62 (in Russian)
[8] Mayer I, Diab H, Reinen D and Albrecht C 1993 *J. Mater. Sci.* **28** 2428
[9] Mayer I, Jacobsohn O, Niazov T, Werckmann J, Iliescu M, Richard-Plouet M, Burghaus O and Reinen D 2003 *Eur. J. Inorg. Chem.* **7** 1445
[10] Pan Y, Chen N, Weil J A and Nilges M J 2002 *Amer. Miner.* **87** 1333
[11] Pon-On W, Meejoo S and Tang I 2008 *Mater. Res. Bull.* **43** 2137
[12] Gafurov M, Biktagirov T, Mamin G, Klimashina E, Putlayev V, Kuznetsova L and Orlinskii S 2015 *Phys. Chem. Chem. Phys.* **17** 20331
[13] Gafurov M R, Biktagirov T B, Mamin G V, Shurtakova D V, Klimashina E S, Putlyaev V I and Orlinskii S B 2016 *Phys. Solid State* **58** 469
[14] Stich T A, Lahiri S, Yeagle G, Dicus M, Brynda M, Gunn A, Aznar C, DeRose V J and Britt R D 2007 *Appl. Magn. Reson.* **31** 321
[15] Chelyshev Y, Gafurov M and Ignatyev I 2016 *BioMed Res. Int.* **2016** 3706280
[16] Galukhin A V, Khelkhal M A, Gerasimov A V, Biktagirov T B, Gafurov M R, Rodionov A A and Orlinskii S B 2016 *Energy Fuels* **30** 7731
[17] Stoll S and Schweiger A 2006 *J. Magn. Reson.* **178** 42
[18] Klyava Y G 1988 *EPR spectroscopy of disordered solids* (Riga: Zinatne) p 320 (in Russian)
[19] Aminov L K, Gafurov M R, Korableva S L et al. 2017 *Phys. Solid State* **59** 564
[20] Schreurs J W H, 1978 *Journ. Chem. Phys.* **69** 2151

IOP Conf. Series: Earth and Environmental Science **155** (2018) 012007 doi:10.1088/1755-1315/155/1/012007

Low-temperature thermal decomposition of heavy petroleum distillates: interconnection between the electrical properties and concentration of paramagnetic centres

M Dolomatov[1,2], M Gafurov[3], A Rodionov[3], G Mamin[3], L. Miquel González[3], A Vakhin[3], A Petrov[2,4,*], R Bakhtizin[1], I Khairudinov[2,4], and S Orlinskii[3]

[1] Bashkir State University, 32 Validy Str., Ufa, Russia
[2] Ufa State Petroleum Technological University, 1 Kosmonavtov Str., Ufa, Russia
[3] Kazan Federal University, 18 Kremlevskaya Str., Kazan, Russia
[4] Institute of Petroleum Refining and Petrochemistry, 12 Iniciativnaya Str., Ufa, Russia

E-mail: petrovanchemistry@gmail.com

Abstract. Changes of paramagnetic centers (PC) concentration in petroleum dispersed systems (PDS) are studied in the process of low-temperature thermolysis. Complex investigation of physicochemical, rheological and electrophysical properties of high-boiling oil fractions is performed. Based on the analysis of the experimental results it can be concluded that the PDS under investigation can be regarded as amorphous broadband organic semiconductors for which PC plays a role of dopant. It shows the perspectives of the asphaltenes usage as a basis for the photovoltaic devices.

1. Introduction

Petroleum fluid can be considered as a petroleum disperse system (PDS) - a hybrid of a solution and a colloidal dispersion (of crystallizing waxes, self-associating asphaltenes, etc.) [1]. Understanding of complex behavior of PDS under the thermal treatment is expected to be a challenge. Nevertheless, as it was stressed by Evdokimov et al., relatively simple experiments on only one of the PDS components can shed light on details of phase transformations in such systems [2], [3].

As one of the constituents of PDS, a paramagnetic phase can be the subject of interest: one gram of PDS contains 10^{16}-10^{21} paramagnetic centers (PC, [4]-[6]). The majority of PC is concentrated in the high-molecular PDS components such as asphaltenes, resins and polycyclic aromatic hydrocarbons (PAH). The content of the high-molecular PDS components could reach the values of 45 wt % in native oils and up to 73 wt % in natural asphalts and bitumen [4]. Assuming that asphaltenes and resins have molecular weight of about 1000 amu [4], they could contain up to one unpaired electron per molecule. Obviously, such high concentration of PC should (or could) affect not only the paramagnetic properties of substance but (at least partially) also the electrical qualities of PDS.

Electrical properties of asphaltenes and bitumen are the matter of intensive research for decades that are briefly reviewed in [7], [8]. It was shown that the temperature dependence of the electrical conductivity in some PDS resembles those for the organic broadband semiconductors [7], [9]-[11]. But there is still no detailed description of the process(es) that would adequately describe the thermal generation-recombination of PC in PDS and influence of PC on the electrical properties of PDS [12].

Among numerous additional motivations to investigate the processes of generation-recombination of PC in PDS, the most significant are related to understanding the mechanisms of photodegradation – the PC linewidth and radicals number decreases with sunlight irradiation time of some tenth of hours [13], as well as by the discrepancies in the PC concentrations derived from the conventional and pulsed EPR measurements [14].

2. Materials and Methods

Samples of asphalt and its group components (asphaltenes and maltenes) were investigated. Asphalt (**A**sphalt produced by **P**ropane **D**easphalting, APD) was obtained by means of second stage propane deasphaltization of the Western Siberian oil tar. The APD was fractionated in two steps. Firstly, asphaltenes and resins were precipitated by adding an excess (1:40) of cold isooctane (Sigma-Aldrich) followed by flushing in Soxhlet's apparatus with hot isooctane. The remaining residue was fractionated into asphaltenes and maltenes by passing the alcohol-benzene mixture (5:1) through a column. Endpoint maltenes fraction was isolated by evaporation of the solvent from the extract. The separation scheme is shown in Figure 1.

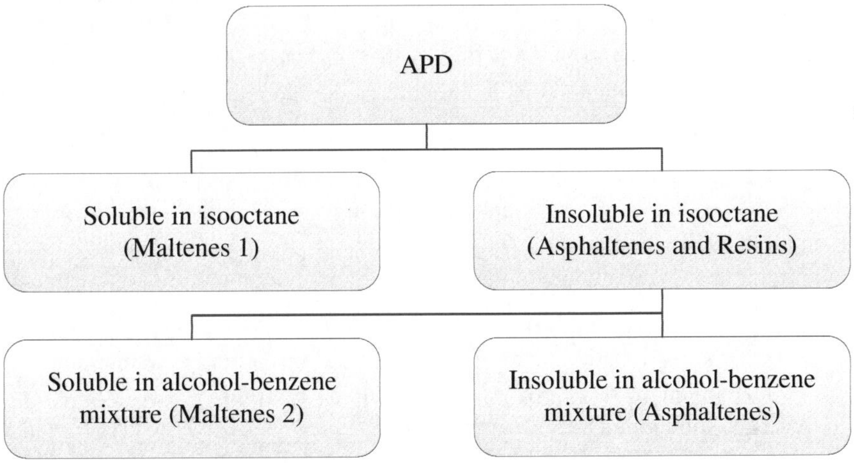

Figure 1. APD separation scheme.

The APD group composition was determined from SARA (Saturated and Aromatic hydrocarbons, Resins and Asphaltenes) analysis using a Gradient-M chromatograph (Ufa, Russia). The measurements of the dynamic viscosity η were performed using a rotary viscometer Brookfield DV2 with a thermostatic cell in the temperature range of (20 – 160) °C. The deviation of temperature during the measurements did not exceed ± 0.5 °C. Elemental analysis was done by X-ray fluorescence measurements (Spectroscan Max GV, S.-Petersburg, Russia). The molecular weight was estimated by naphthalene cryoscopy.

Fourier-transform infrared (FTIR) spectra of asphaltenes and resins were detected using a Shimadzu IRAffinity-1S spectrophotometer in (400-4000) cm^{-1} range with the resolution of 1 cm^{-1}. Samples for IR analysis were prepared as follows. A small amount of the substance was triturated in an agate mortar, poured with hot toluene and thoroughly mixed. The resulting suspension was put on the plate and placed under the IR lamp to vaporize the toluene and dry the film.

Effective electron affinity (EEA) and the effective ionization potential (EIP) were determined from the analysis of the ultraviolet and visible absorption spectra on SF 2000 (Russia) spectrophotometer in the range of (190-780) nm with 1 nm step via the procedure described in [15].

Electrical measurements were performed as follows. Samples are placed into the cylindrical thermostatic cell consisting of a hollow bronze shell and fluoroplastic cover with the mounted aluminum electrode. The shell's bottom is utilized as the second electrode. Homogeneous heating of the measuring cell was provided by the oil-to-oil thermostat, boiling point of the transformer oil is 220 °C. Yokogawa UT750 controller with thermocouple and digital display maintained the required heating rate and accuracy. The electric conductivity was measured in the temperature range of 25 – 165 °C with 20 °C increment. The species were kept at the required temperature during 30 minutes before the measurements' start to avoid adverse inertial effects. The measurements were done only for APD and maltenes due to the considerable complexity of accurate electrical measurements for solid asphaltenes.

EPR spectra of the samples were obtained using ESP-300 spectrometer (Bruker) operating at 9.4 – 9.9 GHz (X-band) in continuous wave mode. Spectrometer's magnetic field can be varied in the range of (20-1600) mT with an accuracy of not less than 0.01 mT, modulation frequency – 100 kHz. PC concentration was estimated at room temperature in the double cavity ER4105DR by comparing the integrated intensities of the spectra of a test sample and a reference samples (Cu-DETC solution and DPPH powder) with a known concentration of PC. Samples heating was occurred in a stream of nitrogen blown through the quartz insert in the high-temperature resonator ER 4114HT equipped with the VT thermostatic system (Oxford Instruments). For each sample the temperature dependence of the EPR spectra in the range of (20 – 500) °C was obtained with (15 – 20) °C increment, heating rate 2 °C/min and temperature setting accuracy of not less than ± 2 °C ([16], [17]) Modulation amplitude 0.1 mT was chosen to investigate narrow EPR lines of free radicals. Microwave power level was equal to 1 mW in order to avoid saturation effects.

3. Results and Discussion.

Investigated APD contains saturated fraction of 8.5% by weight, aromatics – 58.6 wt %, resins – 30.4 wt % and asphaltenes – 17.8 wt %. Some physical and chemical properties for the investigated species are presented in Table 1.

Table 1. Physical properties of the investigated PDS species at 20 °C.

	APD	Asphaltenes	Maltenes
Density at 20 °C (kg/m^3)	1044	1084	978
Average molecular weight, amu	878	1250	669
Elemental composition, wt %			
C	87.0	81.0	84.2
H	10.4	7.7	10.8
S	1.6	4.8	2.2
N	0.30	1.6	0.5
O	0.60	4.9	2.3
Concentration of paramagnetic centers at RT, 10^{18} spin/g			
FR	1.5±0.3	9.0±1.2	4.1±0.5
VP	5.2±0.8	12.8±3.0	5.6±0.1
Electronic characteristics, eV			
EIP	6.18	5.47	6.41
EEA	2.12	1.78	2.04

Table 2 shows the values for the APD viscosity with temperature. Different models for the explanation of the temperature dependencies of η in PDS can be found in literature [18]. Our data can be adequately described by exponent based on the Arrhenius equation

$$\eta = \eta_0 e^{\frac{E_a}{RT}} \qquad (1)$$

where $\eta_0 = (3.0\pm1.2)\cdot10^{-12}$– Pa·s, R is the ideal gas constant and $E_a = (86.1\pm0.9)$ kJ/mol is an activation energy for PDS viscous flow [19].

Table 2. Temperature dependence of the dynamic viscosity of APD

Temperature, °C	Dynamic viscosity (η), Pa·s
60	158.8
70	60.1
80	26.4
90	12.8
100	8.3
120	6.7
140	5.8
160	5.1

FTIR spectrum of asphaltenes extracted from APD is shown in Figure 2. Typical for the asphaltenes peaks are observed in the vicinity of $\approx$ 1600 cm^{-1} (stretching vibrations of the C = C bonds in the aromatic rings), $\approx$ 2850, $\approx$ 2920, $\approx$ 1460 and $\approx$ 1375 cm^{-1} (corresponding to the CH methyl and methylene groups stretching and deformation vibrations). The absorption bands at $\approx$ 803, $\approx$ 866, $\approx$ 694 cm^{-1} correspond to the strong substituted aromatics; the presence of alkyl substituents is proved by the peaks at $\approx$ 720 and $\approx$ 745 cm^{-1}. The absorption bands at $\approx$ 694, $\approx$1142 and $\approx$ 1180 cm^{-1} presumably correspond to OH groups in the tertiary and phenolic position as shown in ref. [20], [21]. Broad absorption band at $\approx$ 3432 cm^{-1} corresponds to the intra- and intermolecular hydrogen bonds, while at $\approx$ 1030 cm^{-1} (most likely) to the sulfoxide group. The intensive peak at $\approx$ 2360 cm^{-1} can be erroneously interpreted as due to the thiol functional group but it more likely corresponds to atmospheric CO_2 impurities, as the adjacent peak at $\approx$ 2342 cm^{-1} confirms ([20], [21]).

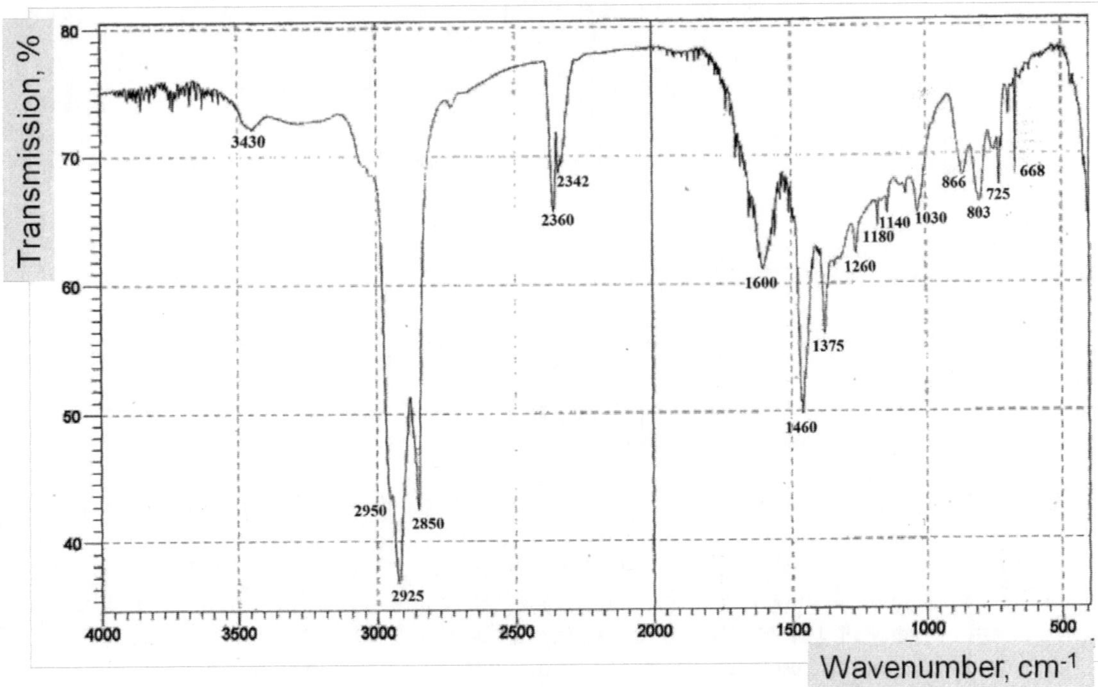

Figure 2. FTIR spectrum of asphaltenes.

The existence of PC is caused by the presence of d-metals (mainly V, Ni, Fe), delocalized π-electrons of the aromatic rings and stable free radicals in PDS. EPR patterns in all the studied samples are very similar. Figure 3 presents that for APD. The EPR spectra originate from the paramagnetic vanadyl complex VO^{2+} in porhyrins ($^{51}V^{4+}$, $3d^1$, electronic spin $S = 1/2$, nuclear spin $I = 7/2$) which will be denoted as VP and from the carbon free radicals (FR) (see [4], [5], [14], [22]). VP molecule is nearly planar defining thus the *g*-factor and hyperfine *A* tensor of axial symmetry. The powder EPR spectrum of VP consists of the 16 "lines" representing the 2×8 hyperfine patterns (the projection of *I* is allowed to take 8 values: ±7/2; ±5/2; ±3/2; ±1/2) for the parallel and the perpendicular complex orientations. The parameters of EPR spectra at RT are $g_{\|} = 1.964 \pm 0.001$, $g_{\perp} = 1.984 \pm 0.001$, $A_{\|} = 16.8 \pm 1.2$ mT, $A_{\perp} = 5.8 \pm 0.8$ mT [23]. The wide dispersion of EPR parameters are caused by the diversity of different structural classes of petroleum porphyrins in PDS that can contain different types of substitutions and binding moieties [22].

Detailed description of VP spectrum is given in refs. [14], [22], [24]. The changes of VP EPR spectra with temperature (pronounced at $T > 200$ °C) were investigated in details in [17]. As it follows from the present experiments, the intensity of VP EPR lines in the temperature range (20-200) °C practically obeys Curie law [4]. Accurate estimation of integral intensity of VP EPR spectrum at higher temperatures is tricky due to the mentioned EPR spectra transformation and is not done in the present research.

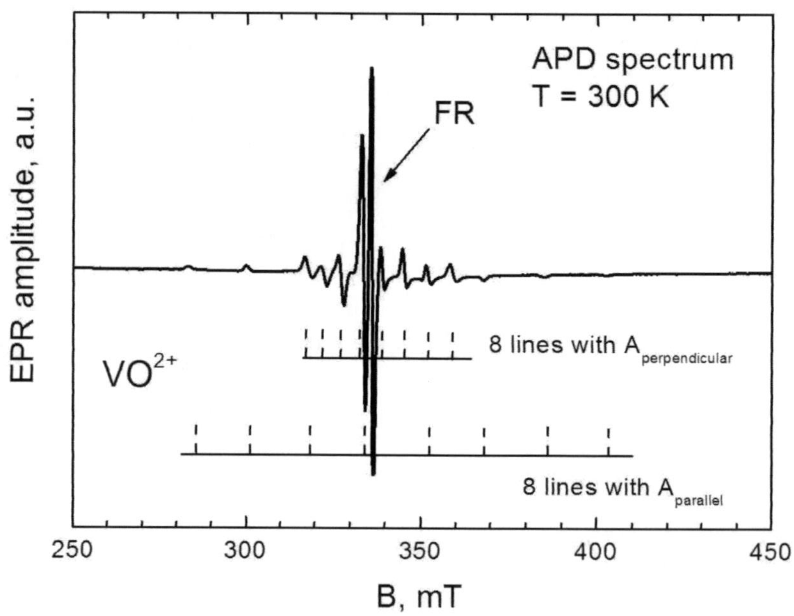

Figure 3. EPR spectrum of APD at T = 300 K. The lines belonging to FR and VP are marked.

The EPR spectrum of FR is a single line with the linewidth of $\Delta H_{1/2} = (0.52\text{-}0.58)$ mT situated at *g*-factor 2.0028 ± 0.0005 (Figure 4). FR lineshape is neither Gaussian (G) nor Lorentzian (L) but a mix of those in ratio of about L/G ≈ 60/40 with the same values of *g*-factors. It is usually assumed that FR in PDS are mainly concentrated in asphaltenes and due to the delocalized π-electrons of the aromatic rings and stable organic radiacals of the side chains [4]. The variations of EPR parameters can be caused by the diversity of different types of unresolved in EPR FRs some of which are anisotropic [6]. While the FR linewidth, lineshape and line position is all investigated fractions do not vary with temperature within the accuracy of our measurements, the intensity I of FR increases non-monotonically (Figure 5). To visualize the $I(T)$ dependencies without Curie contributions, the abscissa and ordinate axis are chosen to be $1000/T$ and $(I \cdot T)$, correspondingly. In the case of pure Curie contribution, the $(I \cdot T)$ would be parallel to the abscissa.

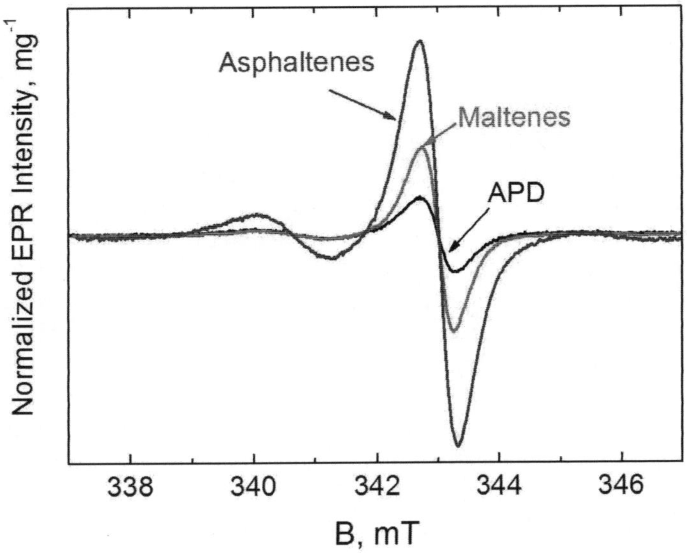

Figure 4. EPR spectra of FR in the investigated samples at $T = 300$ K normalized to the sample masses.

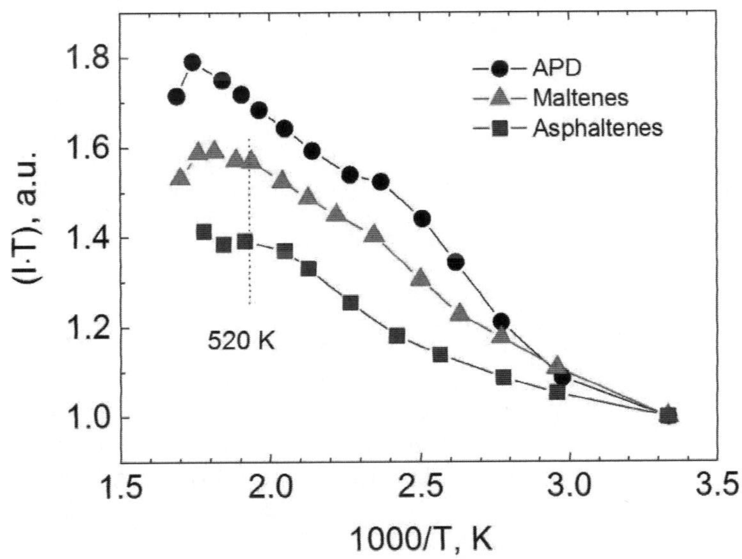

Figure 5. Dependencies of FR relative EPR intensities multiplied by the absolute temperature on the reverse absolute temperature. A dashed line corresponds to T = 520 K.

It has to be noted that for coal the lineshape of EPR spectra for FR are conventionally described by superposition of one Gaussian and up to 3 Lorentzian lines with *g*-factors from 2.0025 for π radicals located on aliphatic and/or aromatic hydrocarbons up to 2.0081 for sulfur containing radicals and strong temperature dependences of *g*-factors in the range (2.0025 - 2.0040) and linwidthes $\Delta H_{1/2}$ are sometimes observed (see [25]-[27]). Obviously, the behavior of EPR parameters of samples studied in this work differs significantly from the mentioned above results. It means in its turn that the proposed for coals explanations of thermal changes of some paramagnetic properties could not be just copy-pasted for PDS systems.

Temperature dependence of EPR intensity for "free" radicals in PDS and other organic systems is conventionally explained by thermal activation of the singlet-triplet transition as it follows from [4], [5], [28]. However, our EPR data do not fully obey the model proposed there. Apparently, it is due to the FR generation and recombination. Carbon bonds weakened by conjugation with aromatic rings of asphaltenes and resins can be easily disintegrated, serving as sources of "free" radicals. The reverse process of FR recombination is also possible and should be taken into account. The complex character of the temperature dependence of PC concentration can be ascribed to the existence of different competitive processes. From one side, the viscosity decrease means the higher rate (frequency) for the PC collisions. Therefore, with temperature and viscosity lowering, the recombination rate increases. From other side, the amount of PC increases with temperature rising and the viscosity decrease. It leads to the breaking of the chemical bonds weakened by conjugation with aromatic moieties of asphaltenes. We hope to offer an appropriate kinetic model that describes the mentioned process(es) in our next publications. One of the possible models is proposed in our recent work [27].

Table 3 represents the temperature dependence of the electrical conductivity of the samples. At ambient pressure and $T = 25$ °C the conductivity is less than 10^{-11} $Om^{-1}\cdot m^{-1}$. With the temperature increase in the range of T = (50-180) °C a dramatic raise of conductivity up to 10^{-5} $Om^{-1}\cdot m^{-1}$ is to obtain. Qualitatively, the conductivity measurements correlate with the temperature dependencies of EPR intensities (Figure 5) and with the higher concentrations of PC in maltenes in comparison to APD (Table 1) that define the earlier begin of the "dielectric-semiconductor" phase transition for maltenes and their higher conductivity at the elevated temperatures.

Table 3. The temperature dependence of the electrical conductivity of the samples

Temperature, °C	Electrical conductivity, $10^{-9}\cdot Om^{-1}\cdot m^{-1}$	
	APD	Maltenes
25	<0.01	<0.01
45	<0.01	<0.01
65	0.2	35
85	5	180
105	20	470
125	110	790
145	300	1400
165	750	2800

There are some evidences that the electrical conductivity in PDS is defined by the charge carriers in asphaltenes [7] and "free" radicals participate in the charge transfer [9]. It can be supposed that as for the other high polymers, the PDS conductivity can be at least partially explained by the jump mechanism (see [7], [11], [30]-[33]).

As shown in Table 1, the ionization energy of the samples is low (EIP varies in the range 5.47-6.41 eV, while EEA in the range 1.78-2.12 eV) and comparable to the ionization energy of PAH having more than four aromatic rings i.e. corresponds to the free radical containing fragments [] This suggestion is confirmed by quantum-chemical calculations [34] and by the electrical conductivity measurements at different temperatures [30], [34]-[36]

Following the results and discussion described above, we can conclude that at least the paramagnetic phase of PDS can be regarded as an amorphous organic semiconductor. The increase of the unpaired electrons with temperature leads to the excess of holes and appearance of the *p*-type semiconductivity. Additionally, the PC as donors of electrons may narrow down the value of the dielectric/semiconductor band gap. All these results in the increase of the concentration of charge carries in the asphaltene containing substances. It means that the control of the "free" radical concentration might be used to obtain the asphaltene based semiconductors with the definite semiconducting properties. This feature could be fruitfully exploited for the construction of new asphaltenes' based photovoltaic devices [37].

4. Conclusion

Complex investigation of physicochemical, rheological and electrophysical properties of high-boiling oil fractions is performed. The results confirm the conclusion about the stable free radicals generation in the PDS multicomponent medium with temperature. The both electrical conductivity and intensity of EPR line with temperature could be explained by radicals accumulation and recombination processes. It is suggested that at least the paramagnetic phase of PDS can be considered as an organic broadband semiconductor. It shows the perspectives of the asphaltenes' usage as a substrate for the photovoltaic devices the properties of which can be changed through the manipulation of electron spins concentration.

Acknowledgments
The work was funded by the subsidy of the Russian Government to support the Program of competitive Growth of Kazan Federal University among World's Leading Academic Centers.

References

[1] Syunyaev Z I 1980 *Chemistry and Technology of Fuels and Oils* **16** 484
[2] Evdokimov I N, Eliseev N Yu and Eliseev D Yu 2001 *J. Petrol. Sci. Eng.* **30** 199
[3] Evdokimov I N, Eliseev N Yu and Eliseev D Yu 2004 *Fuel* **83** 897
[4] Yen T F and Chilingarian G V 2000 *Asphaltenes and asphalts.* Vol. 2 *(Developments in petroleum science)* 40B (Elsiever Science) p 622
[5] Yen T F and Chilingarian G V 1994 *Asphaltenes and Asphalts.* Vol. 1 (*Developments in Petroleum Science*) 40A (Amsterdam: Elsiever Science) p 459
[6] Volodin M A, Mamin G V, Izotov V V and Orlinskii S B. 2013 *J. Phys. Conf. Ser.* **478** 012002
[7] Evdokimov I N and Losev A P 2010 *Energy Fuels* **24** 3959
[8] Hosseini A, Zare E, Ayatollahi S, Vargas F M, Chapman W G, Kostarelos K and Taghikhani V 2016 *Fuel* **178** 234
[9] Dolomatov M, Dezortzev S and Shutkova S 2012 *Journal of Materials Science and Engineering B* **2** 151
[10] Grechishkin V S and Ainbinder N E 1967 *Sov. Phys. Usp.* **10** 237
[11] Zhivulin V E, Pesin L A and Ivanov D V 2016 *Physics of the Solid State* **58** 86
[12] Wang W, Ma Y, Li S, Shi J and Teng J 2016 *Energy Fuels* **30** 830
[13] Guedes C L B, Di Mauro E, De Campos A, Mazzochin L F, Bragagnolo G M, De Melo F A and Piccinato M T 2006 *International Journal of Photoenergy* **2006** 48462
[14] Mamin G V, Gafurov M R, Yusupov R V, Gracheva I N, Ganeeva Y M, Yusupova T N and Orlinskii S B 2016 *Energy Fuels* **30** 6942
[15] Dolomatov M Yu, Mukaeva G R and Shulyakovskaya D O 2013 *Journal of Materials Science and Engineering B* **3** 183
[16] Galukhin A V, Khelkhal M A, Gerasimov A V, Biktagirov T B, Gafurov M R, Rodionov A A, Orlinskii S B 2016 *Energy Fuels* **30** 7731

[17] Biktagirov T B, Gafurov M R, Volodin M A, Mamin G V, Rodionov A A, Izotov V V, Vakhin A, Isakov D and Orlinskii S B 2014 *Energy Fuels* **28** 6683
[18] Daucik P, Višnovský J, Ambro J and Hájeková E 2008 *Acta Chimica Slovaca* **1** 43
[19] Ilyin S O and Streletz L A 2018 *Energy Fuels* **32** 268
[20] Li S, Liu C, Que G, Liang W and Zhu Y 1997 *Fuel* **76** 1459
[21] Calemma V, Iwanski P, Nali M, Scotti R and Montanari L 1995 *Energy Fuels* **9** 225
[22] Gilinskaya L G 2008 *J. Struct. Chem.* **49** 245
[23] Gafurov M, Volodin M, Rodionov A, Sorokina A, Dolomatov M, Petrov A, Vakhin A,Mamin G and Orlinskii S 2018 *Journal of Petroleum Science and Engineering* **166** 363
[24] Gracheva I N, Gafurov M R, Mamin G V, Biktagirov T B, Rodionov A A, Galukhin A V and Orlinskii S B 2016 *Magn. Reson. Solids* **18** 16102
[25] Petrakis L and Grandy D W 1978 *Anal. Chem.* **50** 303
[26] Kudynska J and Buckmaster H A 1996 *Fuel* **75** 872
[27] Liu J, Jiang X, Shen J and Zhang H 2014 *Advanced Powder Technology* **25** 916
[28] Niizuma S, Steele C T, Gunning H E and Strausz O P 1977 *Fuel* **56** 249
[29] Dolomatov M Yu, Rodionov A A, Gafurov M R, Petrov A V, Biktagirov T B, Bakhtizin R Z, Makarchikov S O, Khairudinov I Z and Orlinskii S B 2016 *Magn. Reson. Solids* **18** 16101
[30] Syunyaev R Z and Balabin R M 2007 *J. Disp. Sci. Tech.* **28** 419
[31] Konkin A, Singh C R, Orlinskii S, Mamin G, et al. 2016 *Synthetic Metals* **215** 251
[32] Konkin A, Popov A, Ritter U, Orlinskii S, et al. 2016 *Journ. Phys. Chem. C* **120**, 28905
[33] Niklas A, Mardis K, Banks B, et al. 2013 *Phys. Chem. Chem. Phys.* **15** 9562.
[34] Dolomatov M Yu, Shutkova S A and Desortsev S V 2012 *Petroleum Chemistry* **52** 267
[35] Valyavin G, Fryazinov V, Dolomatov M Yu and Artamonova E V 1980 *Chemistry and Technology of Fuels and Oils* **3** 54
[36] Dolomatov M Yu, Shulyakovskaya D O, Mukaeva G R and Paymurzina N K 2012 *Journal of Applied Physics Research* **4** 83
[37] Chianelli R R, Castillo K, Gupta V, Qudah A M, Torres B and Abujnah R E 2016 US Patent 9293266 B2

IOP Conf. Series: Earth and Environmental Science **155** (2018) 012008 doi:10.1088/1755-1315/155/1/012008

Heating of the porous medium in gas hydrates formation process

I Gimaltdinov[1], **M Khasanov**[2], **M Stolpovsky** [1]
[1]Ufa State Petroleum Technological University,
[2]Sterlitamak Branch of Bashkir State University

E-mail: iljas_g@mail.ru

Abstract. This paper examines the mathematical model of gas hydrates formation in the porous medium with an establishment of the extended region of the phase transition. It also studies the influence of source parameters and parameters on the boundary of the medium in the bed.

1. Introduction

Various technological processes in oil and gas industry accompanied by the formation and the deposition of gas hydrates both in the reservoir and the systems of surface and subsurface equipment of oil and gas fields. Currently, most of the time gas hydrate formation is having negative effect such as emergency stops caused by plants «sclerosis» and also following reconstruction costs. However, hydrate formation may have useful applications such as increasing capacity of underground tanks for carbon dioxide storage. This possibility is connected with sharp growth of gas by weight in the gas hydrate per unit of volume in the compassion with the free state of gas at the same temperature and pressure. [8].

Processes of hydrate formation are accompanied by latent heat release of hydrate formation that could lead to the significant raise of temperature in the porous medium [9]. This effect can be used for cleaning of the porous medium from plaque or for dilution of hydrocarbon systems, which are initially solid for subsequent extraction from subsurface.

The mathematical model of gas hydration and decomposition in the porous medium of an infinite length is examined in papers [1-3, 5-7, 10]. There were obtained automodel decisions and it were shown that depending on the initial and boundary conditions formation and decomposition of the gas hydrate may occur both on frontal surface and extended region. This work describes the mathematical model of formation of gas hydrate in the porous medium of finite-length with an establishment of extended region of the phase transition. There was studied the influence of initial parameters and also parameters on the board of the medium on the evolution of temperature fields.

2. Basic assumptions and equations

Let us consider a linearly-parallel problem of gas hydrate formation in the porous bed of length L ($0 \leq x \leq L$) in case injecting cold gas through its left boundary $x = 0$. The intensity of the gas hydrate formation process is limited by latent heat elimination of hydrate formation. To describe the processes of heat and mass transfer in the porous medium while injecting gas, in the case of gas injection, which are accompanied by gas hydrate formation, we made the following assumptions: the medium porosity is constant, the skeleton of the porous medium, water, and gas hydrate are incompressible and motionless. The assumption about immobility of the liquid is justified by the superiority of the gas filtration velocity over the water filtration velocity with the exception of cases when water saturation is close to one. So in most cases when water saturation of pores is no more than a half, the assumption about its immobility is justified. In addition, we adopt the mono-temperature model of the porous

IOP Conf. Series: Earth and Environmental Science **155** (2018) 012008 doi:10.1088/1755-1315/155/1/012008

medium that is the model where the temperature of the medium and of the saturating substance coincides at all points. The hydrate is a two-species system with the mass fraction of the gas G., and the gas is calorically perfect.

If we orientate the X-axis in the direction of gas flow, system of equations describing the processes of filtration and heat transfer accompanied by gas hydrate formation in the porous medium, which consist of law of conservation of mass and energy; with those assumptions Darcy's law and Ideal gas law looks like [3]:

$$\frac{\partial}{\partial t}\left(mS_g\rho_g + mGS_h\rho_h\right) + \frac{\partial}{\partial x}\left(mS_g\upsilon_g\rho_g\right) = 0,\quad \frac{\partial}{\partial t}\left(mS_l\rho_l + m(1-G)S_h\rho_h\right) = 0,$$

$$\frac{\partial}{\partial t}(\rho c\cdot T) + c_g\rho_g mS_g\upsilon_g\frac{\partial T}{\partial x} = \frac{\partial}{\partial x}\left(\lambda\frac{\partial T}{\partial x}\right) + \frac{\partial}{\partial t}\left(m\rho_h L_h S_h\right),\quad mS_g\upsilon_g = -\frac{k_g}{\mu_g}\frac{\partial p}{\partial x},\quad p = \rho_g R_g T. \quad (1)$$

Here, m is the porosity; ρ_j and S_j ($j = sk, h, l$) are the true density and saturation of the jth phase pores; υ_g, k_g, c_g and μ_g are respectively velocity, permeability, specific heat and dynamic viscosity of the gas phase; p is the pressure; T is the temperature; L_h is the specific heat of hydrate formation; ρc and λ are the specific volume heat and the heat transfer coefficient of the system consisting of the porous medium and gas hydrate, which with those assumptions can be considered as constant values; subscripts h, l, g relates to parameters of hydrate, water and gas respectively

This system is complemented by dependence of the gas permeability coefficient on the gas saturation

$$k_g = k_0 S_g^3,$$

where k_0 – is the absolute permeability of the bed.

When the gas hydrate is formed, regions where the gas, hydrate, and water can be in different states can appear in the porous bed. Following mass and heat balance conditions are being realized on discontinuity surface between regions which have a shock of phase saturation, and also the flows of mass and heat.

$$\left[m\left(\rho_l S_l\left(\upsilon_l - \dot{x}_{(i)}\right) - \rho_h S_h(1-G)\dot{x}_{(i)}\right)\right] = 0,\quad \left[m\left(\rho_g S_g\left(\upsilon_g - \dot{x}_{(i)}\right) - \rho_h S_h G\dot{x}_{(i)}\right)\right] = 0,$$

$$\left[\lambda\frac{\partial T}{\partial x}\right] = \left[m\rho_h L_h S_h \dot{x}_{(i)}\right]. \quad (2)$$

Here $[\psi] = \psi^+ - \psi^-$ – a shock of parameter ψ on interface $x = x_{(i)}$, $\dot{x}_{(i)}$ — velocity of this boundary. We also will consider the temperature and the pressure on these interfaces are assumed to be continuous.

The phase equilibrium condition is adopting in three-phase region where gas, gas hydrate and water present at the same tame.

$$T = T_0 + T_* \ln\left(\frac{p}{p_{s0}}\right). \quad (3)$$

Here T_0 – the initial temperature of the system, p_{s0} – the equilibrium pressure corresponding to initial temperature, T_* – is an empirical parameter depending on the gas hydrate type. It follows from (3) that

$$\frac{\partial T}{\partial x}=\frac{T_*}{p}\frac{\partial p}{\partial x},\quad \frac{\partial^2 T}{\partial x^2}=\frac{T_*}{p}\frac{\partial^2 p}{\partial x^2}-\frac{T_*}{p^2}\left(\frac{\partial p}{\partial x}\right)^2. \tag{4}$$

Let a porous bed of length L at the initial state be saturated by gas and water under the pressure p_0 and the temperature T_0 meeting conditions of the free existing and which are identical in the whole bed:

$$p=p_0,\ T=T_0,\ S_l=S_{l0}\ \ (0\le x\le L,\ t=0). \tag{5}$$

The conditions on the left boundary (x = 0) which is being injected with the p_e and the temperature T_e meeting the conditions of the stable existing of gas and hydrate can be represented in the form:

$$p=p_e,\ T=T_e\ \ (x=0,\ t>0). \tag{6}$$

Let us set the conditions on the right boundary ($x = L$), modeling absence of heat flow though it and the pressure equal to inital:

$$\frac{\partial p}{\partial x}=0,\ \frac{\partial T}{\partial x}=0\ \ (x=L,\ t>0) \tag{7}$$

3. Reduction of the system of equations to a convenient form for numerical solution and analysis of the results

Let us consider the process of hydrate formation in volume region. There are three characteristic regions: the near region where pores are filled by gas and hydrate, intermediate region where gas, hydrate and water are in equilibrium and the distant one which is filled by gas and water. Process of hydrate formation occurs in the intermediate region process. Respectively, two movable surfaces form: between distant and intermediate regions where the process of water to hydrate transformation begins, and between distant and intermediate regions where the process of water to hydrate transformation ends.

Boundary where conditions (6) are valid will be called impermeable.

From (1) the equations of piezoconductivity and thermal conductivity have the form:

$$\frac{\partial}{\partial t}\left(\frac{p}{T}\right)=\frac{k_g}{m\mu_g S_g}\frac{\partial}{\partial x}\left(\frac{p}{T}\cdot\frac{\partial p}{\partial x}\right), \tag{8}$$

$$\frac{\partial T}{\partial t}=\frac{\partial}{\partial x}\left(\kappa^{(T)}\frac{\partial T}{\partial x}\right)+\frac{c_g k_g}{\mu_g \rho c}\frac{p}{R_g T}\frac{\partial p}{\partial x}\frac{\partial T}{\partial x}.$$

The analogical equations for the intermediate region:

$$\frac{\partial}{\partial t}\left(\frac{p}{T}\right)=\frac{k_g}{m\mu_g S_g}\frac{\partial}{\partial x}\left(\frac{p}{T}\frac{\partial p}{\partial x}\right)-\Re\frac{\partial S_h}{\partial t}, \tag{9}$$

$$\frac{\partial T}{\partial t}=\frac{\partial}{\partial x}\left(\kappa^{(T)}\frac{\partial T}{\partial x}\right)+\frac{c_g k_g}{\mu_g \rho c}\frac{p}{R_g T}\frac{\partial p}{\partial x}\frac{\partial T}{\partial x}+\Delta T\frac{\partial S_h}{\partial t},$$

where $\Re=\dfrac{G\rho_h R_g}{1-S_{l0}}$, $\Delta T=\dfrac{m\rho_h L_h}{\rho c}$, $\kappa^{(T)}=\dfrac{\lambda}{\rho c}$ –thermal conductivity coefficient of the bed. If we exclude the time derivative of hydrate saturation from (9) with allowance of equations (5) we will obtain the system for finding the pressure distribution (and relatively the temperature distribution) and hydrate saturation in the extended region:

$$\frac{\partial}{\partial t}\left(\frac{p}{T}\right)=\frac{k_g}{m\mu_g S_g}\frac{\partial}{\partial x}\left(\frac{p}{T}\frac{\partial p}{\partial x}\right)-\frac{\Re}{\Delta T}\frac{\partial T}{\partial t}+\frac{\kappa^{(T)}\Re}{\Delta T}\frac{\partial^2 T}{\partial x^2}+\frac{c_g k_g \Re}{\mu_g \rho c \Delta T R_g}\frac{p}{T}\frac{\partial p}{\partial x}\frac{\partial T}{\partial x}, \tag{10}$$

$$\frac{\partial S_h}{\partial t}=\frac{1}{\Delta T}\left[\frac{\partial T}{\partial t}-\frac{\partial}{\partial x}\left(\kappa^{(T)}\frac{\partial T}{\partial x}\right)-\frac{c_g k_g}{\mu_g \rho c}\frac{p}{R_g T}\frac{\partial p}{\partial x}\frac{\partial T}{\partial x}\right].$$

We will consider that gas injection to the bed on the surface $x = x_{(n)}$ dividing the distant and intermediate region leads to the shock of hydrate saturation from $S^-_{h(n)} = S_{he} > 0$ to $S^+_{h(n)} > 0$ reasoned by the transfer of part of water to hydrate on this boundary. Conditions of mass equilibrium (for gas and water) and energy equilibrium on this surface like it follows from (2) have the form:

$$mS^-_{g(n)}\rho^-_{g(n)}\left(\upsilon^-_{g(n)}-\dot{x}_{(n)}\right)=mS^+_{g(n)}\rho^+_{g(n)}\left(\upsilon^+_{g(n)}-\dot{x}_{(n)}\right)+m\left(S^-_{h(n)}-S^+_{h(n)}\right)\rho_h G\dot{x}_{(n)},$$

$$mS^+_{l(n)}\rho_l\dot{x}_{(n)}=m\left(S^-_{h(n)}-S^+_{h(n)}\right)\rho_h(1-G)\dot{x}_{(n)}, \tag{11}$$

$$\lambda\left(\frac{\partial T}{\partial x}\right)^-_{(n)}-\lambda\left(\frac{\partial T}{\partial x}\right)^+_{(n)}=m\rho_h L_h\left(S^-_{h(n)}-S^+_{h(n)}\right)\dot{x}_{(n)}.$$

Here $\dot{x}_{(n)}$ – is the velocity of motion of the boundary $x = x_{(n)}$; the plus and minus superscripts refer to the values before the interface and behind it. It can be showed that on the surface $x = x_{(d)}$ dividing the intermediate and distant region, the hydrate saturation value is continuous and equal to zero: $S^-_{h(d)} = S^+_{h(d)} = 0$. Then from (2) equations for finding the pressure and the temperature on this boundary have the form:

$$\left(\frac{\partial p}{\partial x}\right)^-_{(d)}=\left(\frac{\partial p}{\partial x}\right)^+_{(d)}, \tag{12}$$

$$\left(\frac{\partial T}{\partial x}\right)^-_{(d)}=\left(\frac{\partial T}{\partial x}\right)^+_{(d)}$$

System of equations (8) and (10) with initial and boundary (5) – (7) conditions on the phase transfer boundary (11) and (12) in the finite-difference form is being solved with using the method of front capturing in nodes of the spatial grid [4].

Fig. 2 shows calculation dependences in hydrate formation in the extended region for the case of impermeable right boundary while gas is being injected to the model bed of length L=2m. with the absolute permeability $k_0 = 10^{-14}$ m^2. There are assumed following values characterizing the system consisting of the porous medium, solid gas hydrate, and gas: $m = 0.1$, $G = 0.12$, $S_{l0} = 0.2$, $\mu_g = 10^{-5}$ Pa·s, $\lambda = 2$ W/(m·K), $L_h = 5\cdot 10^5$ J/kg, $\rho_h = 900$ kg/m^3, $\rho_l = 1000$ kg/m^3, $T_0 = 280$ K, $T_* = 10$ K, $p_0 = 4$ MPa, $p_{s0} = 5.5$ MPa, $c_g = 1560$ J/(kg·K). Figure shows that injection of the cold gas, which is accompanied by hydrate formation, leads to bed heating. Hydrate formation occurs in two. On the first quick stage partial formation in the volume region occurs. On the second stage full hydration of the porous medium occurs. The distant movable region reaches the right boundary of the porous bed as the result. There are two regions exist, the near one contains gas and gas hydrate, the second one contains gas, gas hydrate and water at the thermodynamic equilibrium state. The distribution of hydrate saturation and temperature in three-phase region becomes homogeneous over the time.

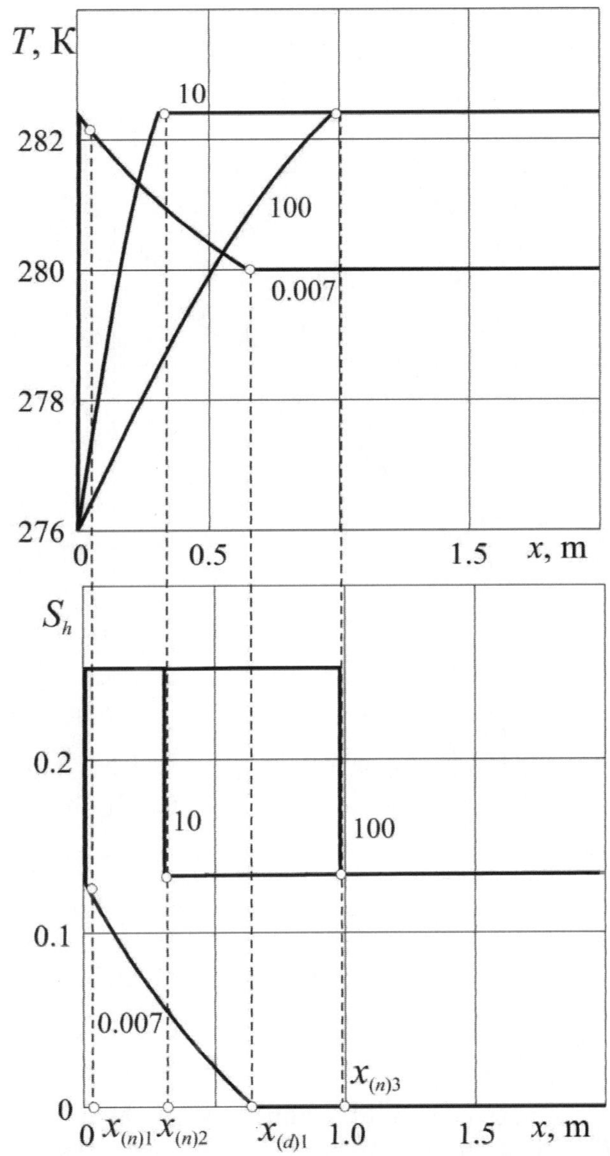

Fig.1. Distribution of temperature and hydrate saturation by X-coordinate in the case of hydrate formation. Values on the curves are the time in hours.

Fig. 1 shows dependence of temperature $T_{(n)}$ on the near boundary from the side of the extended region on the pressure of injecting gas to the bed. Lines 1 and 2 corresponding to values of permeability of the bed $k_0 = 10^{-13}$, $k_0 = 10^{-15}$ m^2. Fig. 2 shows that increasing of the injecting gas pressure and permeability the temperature on the near boundary of hydrate formation becomes higher than the initial temperature of the bed. The reason is that the pressure in the bed is significantly exceeds the equilibrium pressure corresponding to initial temperature T_0. Equilibrium temperature of hydrate formation increases with the growing of pressure. Increasing of pressure of the injecting gas and decreasing of permeability of the bed leads to growing of pressure in the. Therefore the feed of cold gas to the bed leads to its heating.

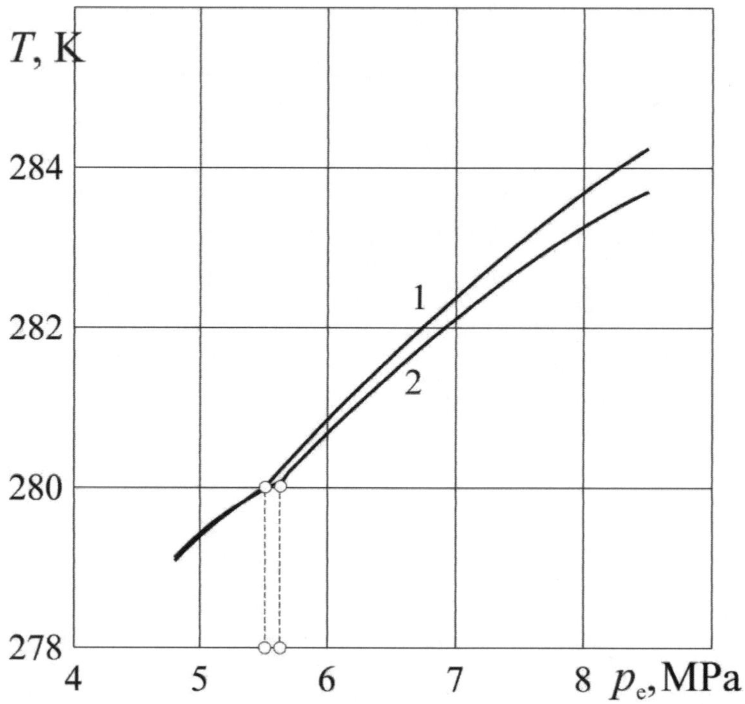

Fig. 2. Distribution of temperature on the near boundary (from the side of extended region) on the pressure of injecting gas: $1 - k_0 = 10^{-13}$ m^2, $2 - k_0 = 10^{-15}$ m^2

4. Conclusion

Numerical simulations of the hydrate formation process in the porous medium with the establishment of the extended region of hydrate have been carried out. It have been shown that injecting of cold gas leads to heating of the bed, and the higher the injecting gas pressure is, the higher heating of the bed.

5. References

[1] Khasanov M K 2015 Investigation of regimes of gas hydrate formation in a porous medium, partially saturated with ice *Thermophysics and Aeromechanics* vol. 22 no 2 pp 245–255

[2] Khasanov M K, Musakaev N G, Gimaltdinov I K 2015 Features of the Decomposition of Gas Hydrates with the Formation of Ice in a Porous Medium *Journal of Engineering Physics and Thermophysics* vol. 88 № 5 pp. 1052–1061

[3] Khasanov M K, Gimaltdinov I K, Stolpovsky M V 2010 Specific features of the formation of gas hydrates during the injection of a cold gas into a porous medium saturated with a gas and water *Theoretical Foundations of Chemical Engineering* vol. 44 no 4 pp. 424-431

[4] V I Vasil'ev, V V Popov, T S Timofeeva 2000 Computational Methods in Development of Oil and Gas Deposits *Izd. Sib. Otd. Ross. Akad. Nauk*

[5] G G Tsipkin 1992 Dissociation of gas hydrates in highly permeable beds *Engineering and Physics Journal* vol.63 no 6 pp. 714-721
[6] G.G. Tsipkin 2001 Mathematical model of dissociation of gas hydrates, coexisting with gas in beds *Rep. Ross. Akad. Nauk* vol. 381 no 1 pp. 56-59
[7] G G Tsipkin 2011 Regimes of dissociation of gas hydrates, coexisting with gas in beds vol. 75 no 5 pp. 24-28
[8] S Sh Byk, Yu F Makogon and V I Fomina 1980 Gas Hydrates *Khimiya (Moscow)*
[9] V A Istomin and V S Yakushev 1992 Gas Hydrates in Nature *Nedra (Moscow)*
[10] V Sh Shagapov and Nadyrova L A 1999 Heating of porous medium, partially filled with gas hydrate, in presence of impermeable boundaries *High Temperature* vol. 37 no 5 pp. 784-789

IOP Conf. Series: Earth and Environmental Science **155** (2018) 012009 doi:10.1088/1755-1315/155/1/012009

Steam-assisted gravity drainage technology enhancement

S Durkin[1] and I Menshikova[1]
[1]Department of geology, oil and gas production and pipeline transportation, Ukhta State Technical University, Ukhta, Russia

E-mail: durkin@bk.ru

Abstract. A hydrodynamic model of a region of Yaregskoye heavy oilfield was build. The results of the simulation have shown that injection capacity along the wellbore of a horizontal well is not uniform. It is determined by the geological heterogeneity of the formation. Therefore, there is importance of enhancing SAGD technology for Yaregskoye oilfield. A new technology was created. The efficiency of the technology is proved by numerical modelling. Horizontal injector and two-wellhead production wells penetrate the formation. Horizontal sections of the wells are located one above the other in the payzone. Wells are divided into two sections. Those sections work simultaneously and independently of one another. This technology allows to increase oil recovery of the oilfield.

1. Introduction

Steam-assisted gravity drainage is one of the most perspective technologies for heavy oil and bitumen production from horizontal wells. In 1968 this process was initiated in Yaregskoye oilfield in thermal mine modification. In modern modification SAGD was developed by R.M.Batler in 1978. Predominantly, nowadays SAGD is used in Canada. In Russia, SAGD is implemented in Komi Republic (Yaregskoye and Usinskoye oilfield) and Republic of Tatarstan (Ashalchinskoye oilfield). [2, 3].

2. Study of SAGD mechanism

For the purposes of the study of SAGD mechanism, a geological model with a help of IRAP RMS of Roxar was built on the section or Yaregskoye oilfield. A sector of the following section was chosen for modelling. Calculations were conducted in hydrodynamic simulator STARS of CMG. Grid block size of the model: 10x10x0.5 m. The number of grid blocks along the horizontal axis: 119x7. The number of vertical layers: 90. Total number of grid blocks: 74970. Permeability varied between 1 and 32000 mD.

The model has 2 horizontal wells. Distance between the horizontal sections of the wells located one above the other – 5 meters. The length of horizontal sections – 1000 meters. The upper well is injector, the lower is producer respectively.

Well constrains:

- Maximum injector BHP: 2500 kPa
- Steam quality: 0.7
- Minimum producer BHP: 600 kPa
- Maximum surface liquid rate: 150 m^3/day

The length of simulation was 12 years. Oil recovery coefficient reached 15.69% and cumulative steam/oil ratio composed 3.8 ton/ton. The results of the simulation have shown that injection capacity along the wellbore of a horizontal well is not uniform. After 5 years of injection, steam moves toward more permeable zones. Actual measurements of temperature and conducted geophysical studies of horizontal production wells have also shown uneven heating of the formation (figure 1).

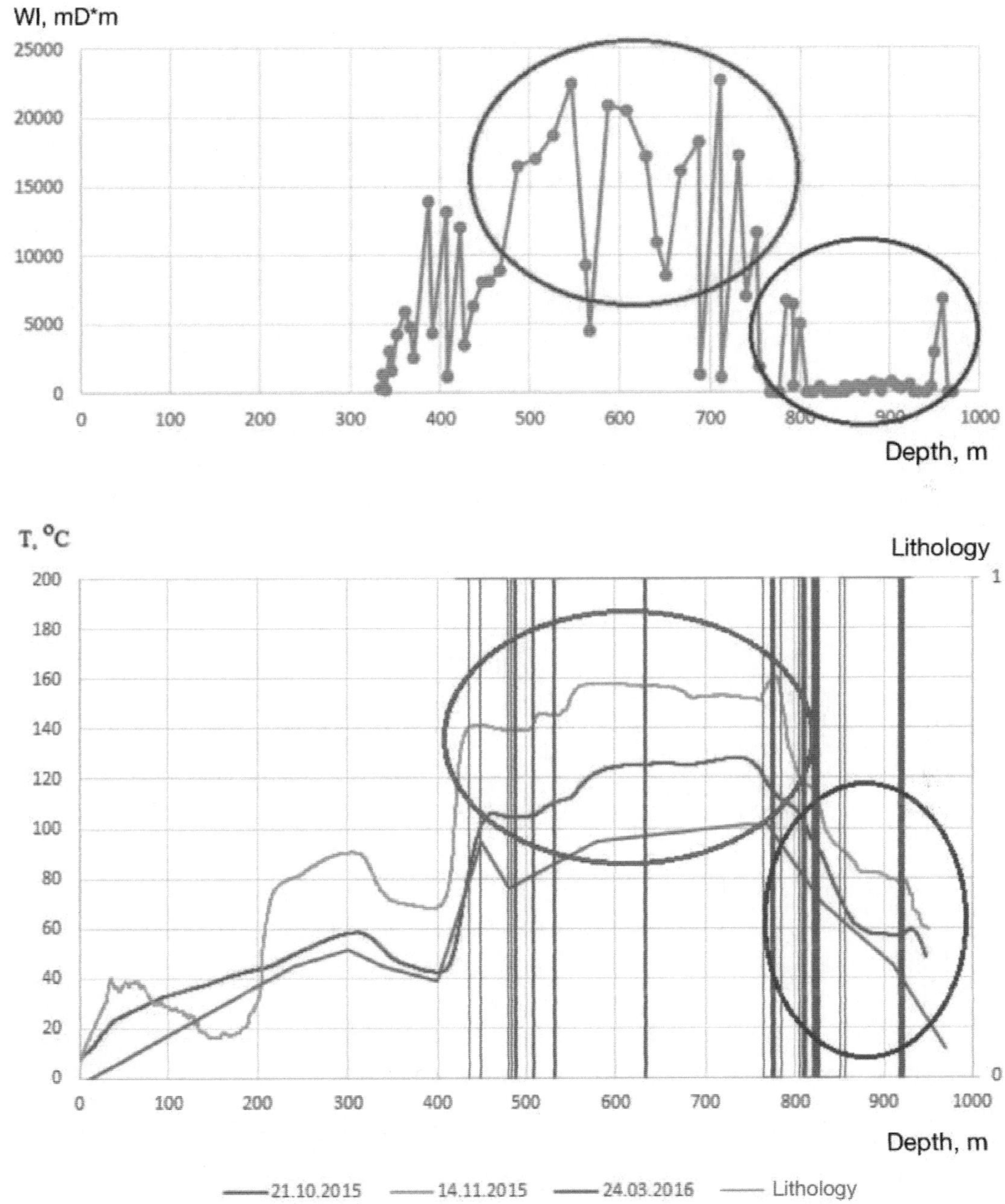

Figure 1 – Calibration of actual temperature measurements with geophysical analysis and WI index in well N.

Results of geophysical studies have shown the distribution of lithology parameter along the wellbore. Value 0 stands for non-collector, value 1 – collector respectively.

Hydrodynamic simulator CMG calculates WI index of the well automatically in every grid block that well penetrates.

Figure 1 shows that high temperature zones coincide with zones having high WI index of the well. Thus, index of the well determines inflow intervals of the reservoir fluid toward horizontal producer wells. Also, it can be noticed that in the zones of heterogeneity (layering) of the formation, development of the steam chamber zone is impeded. SAGD is a very sensitive technology for heterogeneous reservoirs. Thus, there is a high importance of pressure alteration control along the wellbore. Therefore, SAGD technology for conditions of Lyaelskaya sector needs certain optimization.

3. Steam-assisted gravity drainage technology enhancement

In order to enhance SAGD technic for conditions of Yaregskoye oilfield a new technology was developed. The technology is based on differential steam-assisted gravity drainage of the formation (DSAGD).

Technology includes drilling of horizontal injector and two-wellhead producer wells. Horizontal sections of the wells are located one above the other in the payzone. The upper well is injector, lower well is producer respectively. Wells are divided into two sections. Section of the injector is located right above the section of producer. Those sections work simultaneously and independently of one another.

Probable construction of injector well is depicted on figure 2.

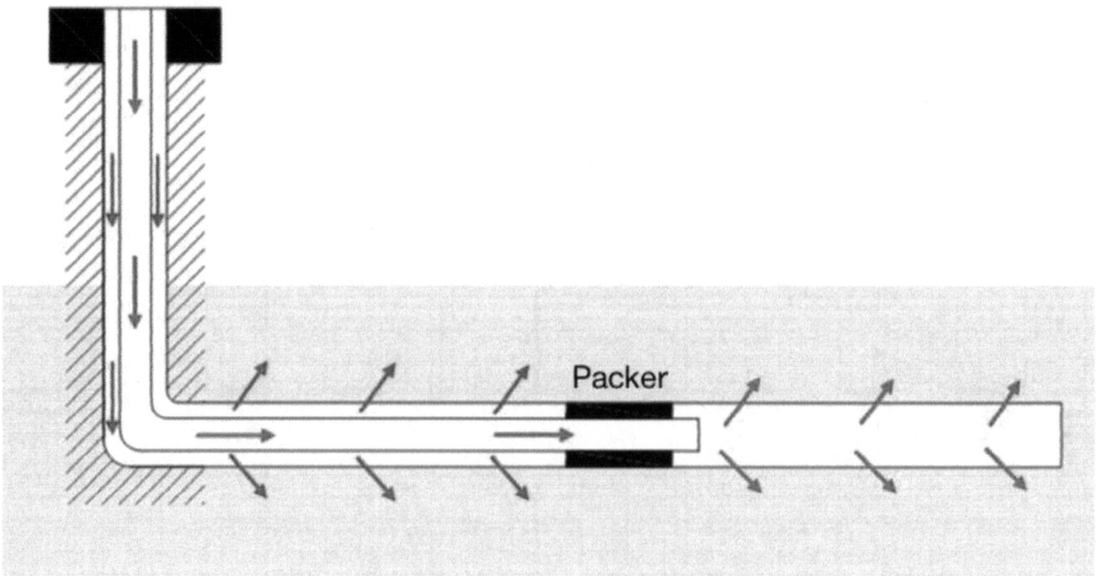

Figure 2 - Probable construction of injector well

In order to divide the sections of producer wells, an installation of a packer in the central part of the horizontal borehole is needed. Oil is produced through both wellheads (figure 3).

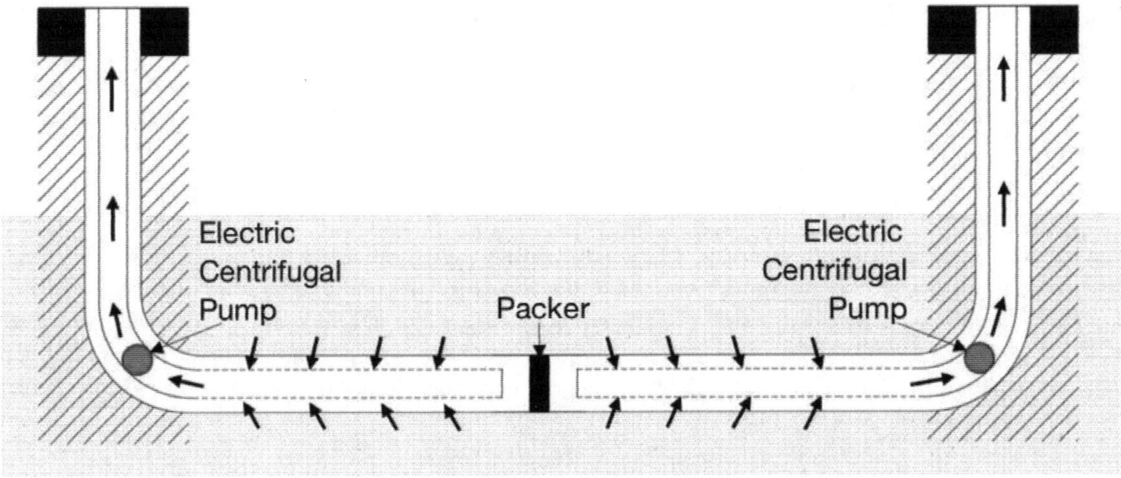

Figure 3 – probable construction of producer well

The following technology allows to increase oil production and enhance thermal sweep efficiency of the formation due to redistribution of the pressure in the formation.

The efficiency of the technology is proved by numerical modelling. For the purposes of the experiment, analogues model of the sector of Yaregskoye oilfield was created. Horizontal wells are divided into 2 sections each having 500 meters in length. The length of modelling composed 12 years as well.

Comparison of the temperature distribution in the formation using SAGD and DSAGD is depicted on figure 4.

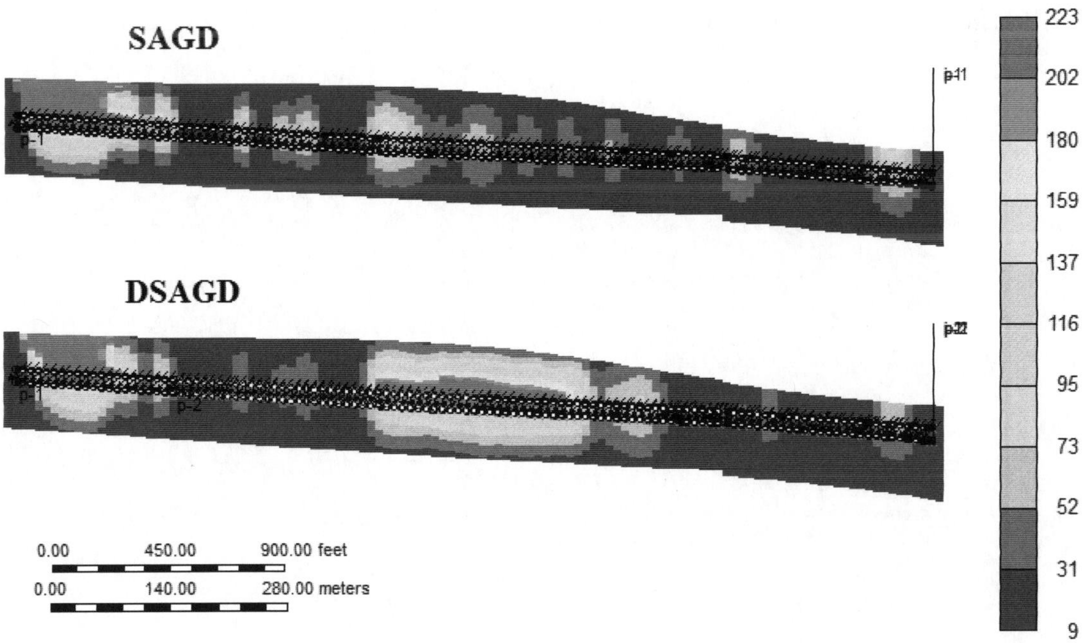

Figure 4 – Temperature distribution in the model after 2 years

Control of process of steam injection into horizontal well allows to enhance thermal sweep efficiency of the formation. Recovery factor using DSAGD reached 28.4%, steam/oil ration composed 2.82 ton/ton.

4. Conclusions

Mechanism of SAGD in conditions of Yaregskoye oilfield shows that the process of steam chamber distribution will depend on geological features of the formation. Actual temperature distribution along the wellbore of production well and conducted geophysical studies also proved the above stated conclusion. In the process of steam injection, steam chamber accumulates in the zones having high WI index. Therefore, the following zones need differential approach. In order to implement this approach, DSAGD technology was developed.

The following technology allows to increase recovery factor and enhance thermal sweep efficiency of the formation due to redistribution of the pressures in the formation. For the rector of Yaregskoye oilfield, the increase of the recovery factor composed more than 12.7% in comparison with conventional technology.

Magnitude of the following increase also depends on the number of sections. Therefore, development of technologies for implementation of smart wells in Yaregskoye oilfield is actual problem.

References

[1] Butler, R.M. Method for continuously producing viscous hydrocar-bons by gravity drainage while injecting heated fluids. US Patent 4344485 A

[2] Maganov, N. U. Expertise of shallow heavy oilfields development [Text] / N. U. Maganov et al// Oil&Gas Journal Russia. – 2015. - №5. – P. 60-63.

[3] Technological principles of development of abnormally viscous oilfields and bitumens. / L. M. Ruzin et al - 2 edition, revised. - Moscow ; Izhevsk : Institute of computer studies, 2015. - 480 p.

IOP Conf. Series: Earth and Environmental Science **155** (2018) 012010 doi:10.1088/1755-1315/155/1/012010

Changes in the composition and properties of Ashalchinskoye bitumen-saturated sandstones when exposed to water vapor

E Korolev, A Eskin, A Kolchugin, V Morozov, M Khramchenkov and R Gabdelvalieva

Institute of Geology and Petroleum Technologies, Kazan Federal University, Kazan, Russia

E-mail: eskin.aleksey@gmail.com

Abstract. Ashalchinskoye bitumen deposit is an experimental platform for testing technology of high-viscosity oil extraction from reservoir rocks. Last time for enhanced of oil recovery in reservoir used pressurization a water vapor with a temperature of ~ 180 ° C (SAGD technology). However, what happens in sandstone reservoir is little known. We did a study of the effects of water vapor on the structural components of bitumen saturated sandstone. In paper were studied the rock samples at base condition and after one week exposure by water vapour. The thermal analysis showed that steaming helps to removes light and middle oil fractions with a boiling point up to 360 ° C from oil saturated sandstones. Content of heavy oil fractions virtually unchanged. Studying the composition of water extractions of samples showed that the process of aquathermolysis of oil is accompanied by a lowering of the pH of the pore solution from 7.4 to 6.5 and rise content in several times of mobile cations Ca^{2+}, Mg^{2+} and HCO_3^-, SO_4^{2-} anions. Follows from this that the thermal steam effect by bitumen saturated sandstones leads to partial oxidation of hydrocarbons with to form a carbon dioxide. The source of sulfate ions were oxidized pyrite aggregates. Due to the increasing acidity of condensed water, which fills the pore space of samples, pore fluid becomes aggressive to calcite and dolomite cement of bitumen saturated sandstones. As a result of the dissolution of carbonate cement the pore fluid enriched by calcium and magnesium cations. Clearly, that the process is accompanied by reduction of contact strength between fragments of minerals and rocks. Resulting part of compounds is separated from the outer side of samples and falls to bottom of water vapor container. Decreasing the amount of calcite and dolomite anions in samples in a steam-treated influence is confirmed by X-Ray analysis. X-Ray analysis data of study adscititious component of rocks showed that when influenced of water vapor to bitumen saturated sandstones there are the processes of transformation of clay minerals. Mixed-illite-montmorillonite phase is primarily exposed to changes. In this case we fix initial stage of the destruction of polycrystalline particles mixed-mineral. Reducing the size of clay minerals particles along the normal to the layers (L_{001}) is the result from lower energy costs of delamination and disintegration the crystallites along this direction, in comparison with others. Thus, using of SAGD technology at exploitation of Ashalchinskoye bitumen saturated reservoir will be followed by acidification of the pore fluid, activation processes of dissolution calcite cement and transformation of mixed-layers clays minerals.

1. Introduction

Ashalchinsky bitumen field is one of the experimental objects for testing and optimization of high-viscosity oil extraction technologies located in the territory of the Republic of Tatarstan [1]. In the tectonic, the field is located on the western side of the Southern Tatar arch [2] (fig. 1). The productive strata are localized in the sandstones of the Sheshminsky horizon of the Ufimian of the Lower Permian system [3]. The cap rocks are the "lingual" clays of the Baytugan horizon of the Kazanian of the Middle Permian system (fig. 2). Bituminous sandstone layers are characterized by variable reservoir properties and heterogeneous hydrocarbon saturation. The highest values of the capacitance-filtration parameters and the residual oil saturation are noted in the arched parts of the low-amplitude brachianticlinal structures, the smallest ones in the saddles between them. Within the Ashalchinsky field, the deposits of high-viscosity oil are located at relatively shallow depths from the Earth's surface, from 80 to 200 m [5, 7]. It is believed that they formed as a result of the ascending migration of water-oil fluids from the underlying oil-saturated succession localized in Carboniferous successions [4, 6].

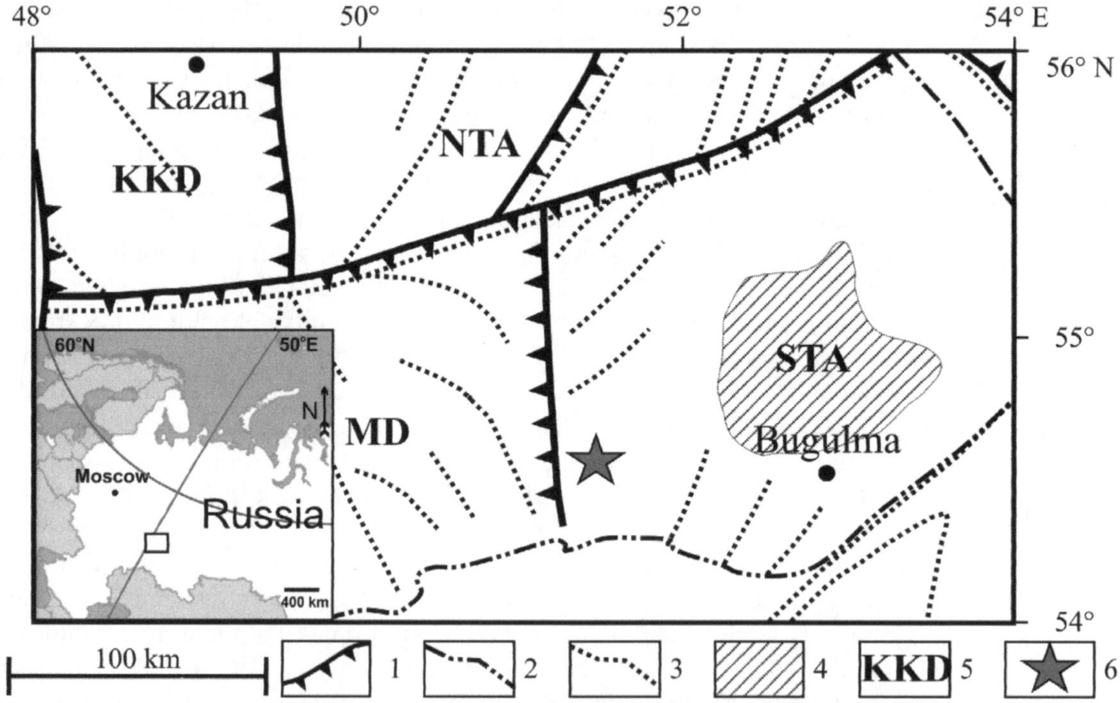

Figure 1. Object setting on schematic on tectonic map.
Legend: 1 – regional fractured zone; 2 – structural boundary; 3 – fractured zone; 4 –arch structure; 5 – name of structural elements (KKD – Kamsko-Kinelsky depression; NTA – North-Tatar arch; STA – South-Tatar arch; MD – Melekesskaya

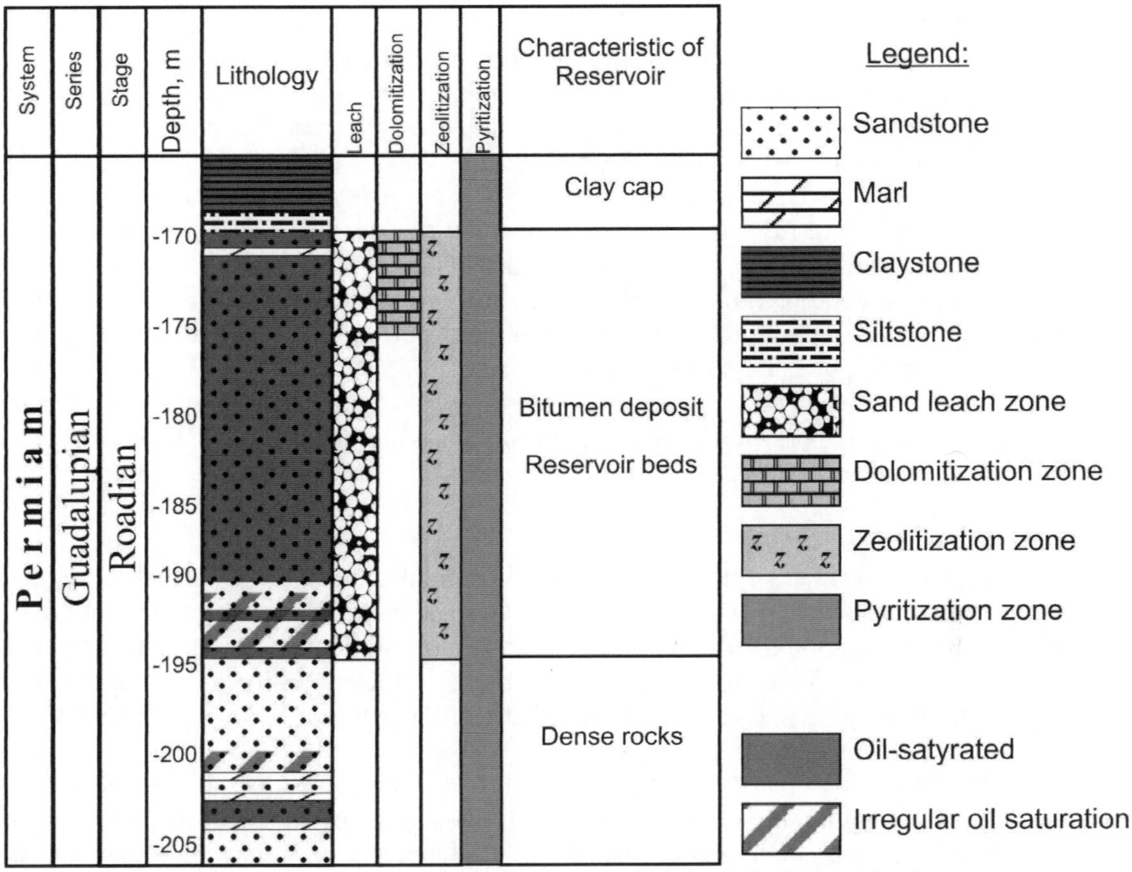

Figure 2. Schematic geological section of reservoir facies with diagenetic changes

Ashalchinskoye bitumen deposit is an experimental platform for testing technology of high-viscosity oil extraction from reservoir rocks. Last time for enhanced of oil recovery in reservoir used pressurization a water vapor with a temperature of ~ 180 ° C and pressure from 0,4-0,55 to 2,5-4,0 MPa (SAGD technology) [10]. However, what happens in sandstone reservoir is little known.

The SAGD technology has proven itself in the development of bituminous sandstones in Canada. The similar method of hydrocarbon extraction was applied on the Ashalchinskoye bitumen field. In the producing formation sandstone zone with thickness of 15-25 m were drilled two horizontal wells, located one above the other. In the upper intake well is supplied water vapor with a temperature of ~ 180 ° C, which heats the bituminous sandstone bed. Condensing water is enriched by hydrocarbons and enters to the lower producing well. Thus, between the two wells going the circulation of the water-oil fluid. The formation of intensive circulation zone within a local area can lead to illuviation not only hydrocarbons from sand reservoirs but also their mineral components. If it poorly consolidated sandstones, then it can lead to problems in their development. Based on this concept the modeling of the effect of water vapor on the structural components of the bituminous sandstone of the Ashalchinskoye deposit was carried out.

2. Methodology

To assess the given fuel type adequacy for thermal energy production it is necessary to know its thermal characteristics and blend composition. The mentioned combined are to determine the combustible shale effectiveness as a fuel resource. In consideration of this the main research methods were X-ray, thermal and SEM analyses.

X-ray phase analysis. The X-ray phase analysis was performed on the diffractometer D2 Phaser (Bruker, Germany) used for measuring the powder products in the Bragg-Brentano geometry, with the use of monochromatic CuKα-radiation (λ = 1.54178 Å), in the step-scan mode. Measuring and recording modes: X-ray tube voltage-30 kV, current-30 mA. Scanning step-0.02°. Speed-1 deg/min. The range of scanning angles in the Bragg-Brentano geometry 3-40°. The standard powder compositions were used.

The thermal analysis was performed on the device STA 449 Jupiter F3 to determine the calorific value and phase transitions of the combustible shale structural components. The burning ranged from 30 to 1000 ° C, the heating step – 10 deg. / min. with constant air blowing.

3. Results and Discussions

Lithological studies of terrigenous bitumen-saturated rocks are showed that the sandstone layer in both the dense and expanded bitumen-saturated part is characterized by a homogeneous composition of the allotigenic component. The prevalent are fain-grained fragments of siliceous and effusive rocks (~ 70%), in lesser quantities there are subangularity grains of quartz (~ 15%), angular grains of albite (~ 7%) and microcline (~ 3%), chlorite flakes (~ 3%), Rare rounded fragments of carbonate rocks (~ 2%).

Optical-microscopic data and sieve analysis of the sandstones fractional composition showed predomination of 0.1-0.25 mm size grains. Grains of siltstone and clay size are in a subordinate amount. At the same time, there is a tendency to reduce the content of silstone component in the productive part of the sandstone pack in comparison with similar unchanged sandstones. The fragments are cemented by argillo-calcareous cement. In dense sandstones, the share of cement is 15-35%, in bituminous-saturated - 3-12% as shown in Table 1.

In intervals of bitumen-saturated sandstones are presence higher porosity and uniform saturation by oil hydrocarbons. The increase of the void space volume was due to the partial leaching of calcite from sandstone cement which substantially reduced the carbonate content of the rocks. Obviously, the first stage of the fluid impact was associated with the advent of aggressive solutions in the sandstone bed which dissolved the calcite cement and thus increased their filtration-capacitive properties. Aggressiveness of solutions to carbonates was probably due to the high content of carbon acid in them. The source of gas was the oxidizing oil deposits of Carboniferous deposits.

Table 1. Fractional composition, carbonate content and porosity of sandstones

Rock type	Fraction content, %					Carbonate content, %	porosity, %
	0.5-0.25	0.25-0.1	0.1-0.05	0.05-0.01	>0.01		
Dense sandstone, unchanged	0.14	61.43	18.43	14.07	5.93	15.66	6.16
	2.91	66.69	17.68	8.92	3.80	28.36	4.98
	0.0	61.32	18.82	10.62	9.24	33.39	3.44
	0.26	65.92	18.63	10.91	4.28	33.96	3.16
	0.46	64.55	18.91	13.97	2.11	35.45	2.31

IOP Conf. Series: Earth and Environmental Science **155** (2018) 012010 doi:10.1088/1755-1315/155/1/012010

	2.07	87.47	2.02	4.04	4.40	10.73	7.16
Bitumen-	4.90	75.51	7.35	9.08	3.16	6.71	8.85
saturated	1.50	87.51	4.65	3.75	2.59	4.01	10.98
sandstone	7.52	70.35	8.06	9.38	4.96	3.68	12.81
	6.42	77.51	5.22	5.50	5.35	3.12	13.42

X-Ray analysis data of study adscititious component of rocks showed that when influenced of water vapor to bitumen saturated sandstones there are the processes of transformation of clay minerals. Mixed-illite-montmorillonite phase is primarily exposed to changes. The diffraction data showed as the duration of the water vapor processing on sample are changing profile of basal diagnostic reflections from flat grid [001] d001 = 14,32 Å. This is showed in reduction of peak reflection intensity, its widening and asymmetry towards small angles. The decrease of the peak intensity and its widening are the result of reducing the size of the crystallites in the polycrystalline structure of the mixed-phase. In this case we fix initial stage of the destruction of polycrystalline particles mixed-mineral. Reducing the size of clay minerals particles along the normal to the layers (L001) is the result from lower energy costs of delamination and disintegration the crystallites along this direction, in comparison with others.

It was found that water vapor helps to bitumen viscosity reducer and partially hydrocarbons removal from sandstone pores. Intermediate oil fractions with a boiling point temperature up to 360 °C were completely washed out and amount of residual heavy fractions with a boiling point temperature above 400 °C decreased from 6.0 to 4.0%. X-ray analysis of the samples showed that content of calcite and clay minerals (illite-montmorillonite and chlorite) in bituminous sandstones after treatment with water vapor decreases [7]. Removal from the rock of the main cementing components leads to changes of sandstones physico-mechanical properties. After 50 hours of water-vapor treatment their density decreases from 2.06 to 1.8 g/cm3, unconfined compressive strength from 2.75 to 1.23 MPa, bending strength from 0.11 to 0.05 MPa, bond value from 0.45 to 0.21 MPa. It leads to increases the effective porosity from 10.7 to 18.2%. After 100 hours of water treatment, sandstones are dissolved and becoming sand. Granulometric analysis of sand showed amount reduction of clastic component, clay and silt fractions in composition. The rock became more uniformly, fine-grained.

Experimental work was performed to evaluate the effect of water vapor on the mechanical properties of bituminous sandstones. From the test core of the bituminous sandstone layer, cubes and cylinders were cut out, which were placed in a water bath.

At certain time intervals, the samples were tested for uniaxial compression (Rcomp.) and tension (Rst.). The obtained results presented in Table 2. They showed that the duration of the action of water vapor on bitumen-containing rocks leads to a progressive decrease in their strength characteristics. A similar tendency is due to a progressive decrease of the binding force (C) in the rocks due to the leaching of calcite cement and the elution of the hydrocarbon substance.

Table 2. Mechanical properties of Ashalchinskoye deposit sandstones

Measurement parameters	Time of water vapor action on bituminous sandstones, hours			
	0	50	100	200
	Poorly consolidated bituminous sandstone			
$R_{comp.}$, MPa	2.75	1.25		
$R_{st.}$, MPa	0.11	0.05	break	
C, MPa	0.45	0.21		

Strong cemented bituminous sandstone				
$R_{comp.}$, MPa	36.00	22.50	16.50	15.75
$R_{st.}$, MPa	1.44	0.90	0.66	0.63
C, MPa	5.94	3.71	2.72	2.61

Research data show that the effect of water vapor on rocks can lead to a significant drop in the strength characteristics of sand collectors in the hydrocarbon recovery process (fig.3).

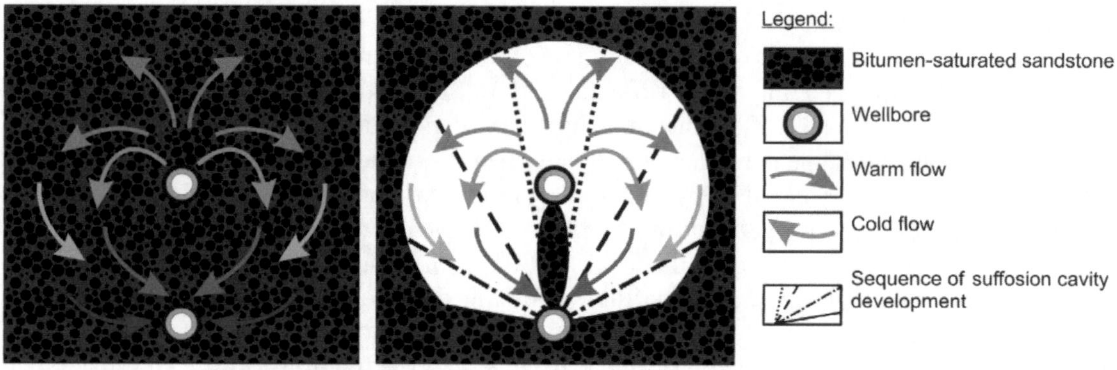

Figure 3. Modeling the development of the suffosive cavity

Estimation of the suffosive cavity size, as well as the time of its formation, according to Boek E.S. [8] is carried out taking into account the physical filtration rate q0 of the fluid in the pores of each rock type, which characterizes the degree of cementation of the rock. The limiting cavity size r_∞ can be calculated from the relation:

$$r_\infty = \frac{Q}{2\pi q_0 m_\infty T}$$

where Q – production well rate; T – linear dimension of exposed by drilling area; m_∞ – maximum achievable value of the suffusion cavity porosity; T – length of the exposed section by horizontal well. The time dynamics of the suffosive cavity formation can be obtained from the relation:

$$m = \frac{q}{q_0}\left[1 - \left(1 - m_0\frac{q_0}{q}\right)exp(-\alpha q_0 t)\right]$$

where q - filtration rate; m0 – initial value of porosity; α – erosion rate of the soil.

Choosing typical values of q0 for different degree of cementation of rocks, we obtain the corresponding values of r_∞.

Figure 4 shows plots of formed suffosion zone size from the degree of rocks binding force for sections of two horizontal wells of 200 and 400 m in length, and a flow rate of 20 and 34 tons per day, respectively. It can be seen from the graphs that when the cement is weakly cemented, the size of the cavity generatrix can reach considerable values.

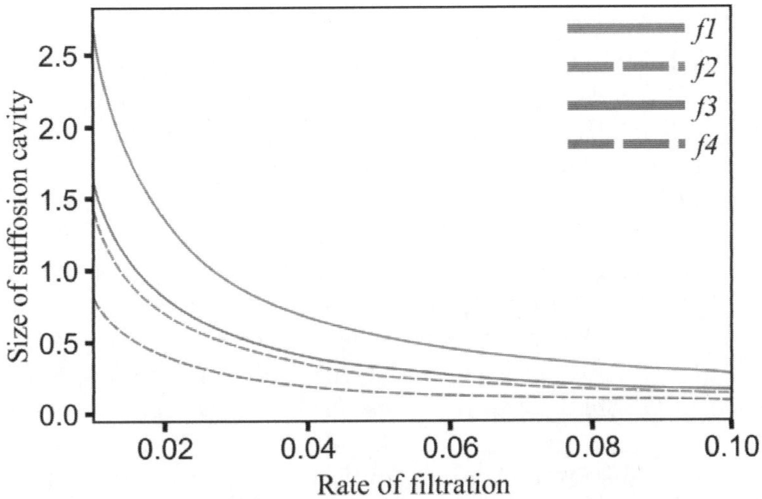

Figure 4. The curves of the dependence of the formed suffusion cavity size in the bitumen deposit from the degree of rocks cementation.

Legend: f1 - flow rate 20 m / day, length of the section 200 m; f2 - flow rate 20 m / day, the length of the section is 400 m; f3 - flow rate 34 m / day, length of the section 200 m; f4 - flow rate 34 m / day, the length of the site is 400 m.

4. Conclusions

Thus, the simulation showed that the process of bituminous sandstones development of the Ashalchinskoye field using the SAGD-technology will be accompanied by loosening of the rock structure and the suffosive removal of mineral particles by a filtration stream into producing wells. To theoretically substantiate of mineral particles removal possibility from the sand layer was calculated rock stability coefficient using the formula for calculated:

$$K = \frac{2\varepsilon H g \rho_r - P_{BH}}{\sigma_{CH}}$$

where K – dimensionless coefficient; H – depth of productive strata; ε – coefficient of rock lateral thrust; g – acceleration of gravity; ρ_r – rock density; P_{BH} – bottom-hole pressure; σ_{CH} – rock compressive strength. Substituting the values corresponding to these quantities, we obtain the variation of K from 15 to 53. It is believed that at K = 1 the rocks are in an equilibrium state, for K> 1 - the rocks are in a stressed state.

When most of the calcite cement was removed from polymictic sandstones, the effects of aggressive solutions began to be exposed to the surfaces of allotigenic grains, primarily the weathered fragments of effusive rocks consisting of albite. The fragile, corroded surface acquired by the wreckage during transportation greatly contributed to the activation of hydrolysis processes on them. According to research [9], the feldspars hydrolysis is accompanied by ion-exchange reactions between H+ and Na+, as a result of which sodium ions leave for solution, and on the plagioclases surface the gibbsite-hydrogibsite layer is formed. Apparently, something similar happened on the surface of fragments of effusive rocks. Before the formation of kaolinite, the process did not reach, which probably contributed to the arrival of water-oil fluids.

Thus, the result of the SAGD-development will be the destruction of the bottom-hole area, the flow of sand into production wells and as a consequence, the colmatage and underground equipment wear.

References:

[1] Muslimov R H 2005 Unconventional oil deposits are a significant potential for the further development of the old oil-producing regions *Georesources*. **1** 2-8

[2] Gatiyatullin N S 2010 Features of the spatial location of Permian bitumens and underlying oil deposits on the territory of the Republic of Tatarstan *Neftegazovaya Geologiya Theory and practice*. **5(3)** 1-12

[3] Hisamov R S 2007 *Geology and development of deposits of natural bitumen of the Republic of Tatarstan* (Kazan, Publishing house "FEN")

[4] Gilyazov T R 2011 Hydrogeological conditions for the location of bitumen deposits in the sediments of the Ufa and Kazan longlines *Georesources*. **37(1)** 41-44

[5] Khisamov R S, Shargorodsky I E and Gatiyatullin N S 2009 *Oil-bitumen content of the Permian deposits of the South Tatar arch and the Melekess depression* (Kazan: Publishing house "FEN")

[6] Khisamov R S, Sukhov K A, Gatiyatullin N S, Voitovich S E and Lieberman V B 2012 On the search for super-viscous oil in the southwestern part of the South Tatar arch *Geology of oil and gas*. **3** 54-58

[7] Korolev E A, Bakhtin A I, Eskin A A and Khanipova R R 2016 Diagenetic changes of sandstone reservoir of Ashalchinskoye bitumen deposit *Neftyanoe khozyaystvo - Oil Industry*. **10** 26-28

[8]Boek E S, Hall C and Tardy P M J 2012 Deep Bed Filtration Modelling of Formation Damage Due to Particulate Invasion from Drilling Fluids *Transp. Porous Med*. **91** 479-508

[9] Shvartsev S L 2007 Basic processes and mechanisms of the water-rock system evolution. *Bulletin of the Tomsk Polytechnic University*. **311**(1) 105-109

[10] Kurochkin B M, Baldenko D F, Rogachev O K and Studenskiy M N 2007 New technologies of heavy crudes and bitumens production at simultaneous drilling on depression *Neftyanoe khozyaystvo - Oil Industry*. **6** 75-82

IOP Conf. Series: Earth and Environmental Science **155** (2018) 012011 doi:10.1088/1755-1315/155/1/012011

Perspective usage estimation of Volga region combustible shale as a power generating fuel alternative

E Korolev[a], E Barieva[b], A Eskin[a]

[a] Institute of geology and petroleum technologies, Kazan Federal University, Kazan, Russia

[b]Kazan State Power-engineering University, Kazan, Russia

e-mail: eskin.aleksey@gmail.com

Abstract. A comprehensive study of combustible shale, common within Tatarstan and Ulyanovsk region, is carried out. The rocks physicochemical parameters are found to meet the power generating fuels requirements. The predictive estimate of ash products properties of combustible shale burning is held.
Minding furnace process technology it is necessary to know mineral and organic components behavior when combustible shale is burnt. Since the first will determine slagging properties of energy raw materials, the second – its calorific value. In consideration of this the main research methods were X-ray, thermal and X-ray fluorescence analyses.
Summing up the obtained results, we can draw to the following conclusions:
1. The combustible shale in Tatarstan and the Ulyanovsk region has predominantly low calorific value (Q_b^d = 5-9 MJ/kg). In order to enhance its efficiency and to reduce cost it is possible to conduct rocks burning together with some other organic or organic mineral power generating fuels.
2. High ash content (A^d = 60-80%) that causes a high external ballast content in shale implies the appropriateness of using this fuel resource next to its exploitation site. The acceptable distance to a consumer will reduce unproductive transportation charges for large ash and moisture masses.
3. The performed fuel ash components characteristics, as well as the yield and volatiles composition allow us to specify the basic parameters for boiler units, designed for the Volga combustible shale burning.
4. The noncombustible residual components composition shows that shale ash can be used in manufacture of materials of construction.

1. Introduction

Coal has always been one of the major power resources for thermal power complexes. Despite gasification attempts solid fuels still have an enormous demand in the industry. However, the changed country's economic doctrine requires new approaches to supply regional thermal power stations and heating plants with fuel resources that use local raw materials [1].

In Tatarstan combustible shale of middle upper Jurassic age can be an alternative to imported coal. In the country's territory eight shale bearing strata are known that are separated from each other and located within Tetyushevo, Buinsk and Drozhzhanovsk areas [1]. Similar rocks can be followed to the

Ulyanovsk region up to the village «Slancevyj rudnik". Estimated reserves by the data type $C_2 + P_1$ are about 900 million tons, in the Ulyanovsk region within Undorsk area – 26.8 million tons. Combustible shale occurs at shallow depth and form aged spreading strata with the power up to 5.0 m (Fig. 1).

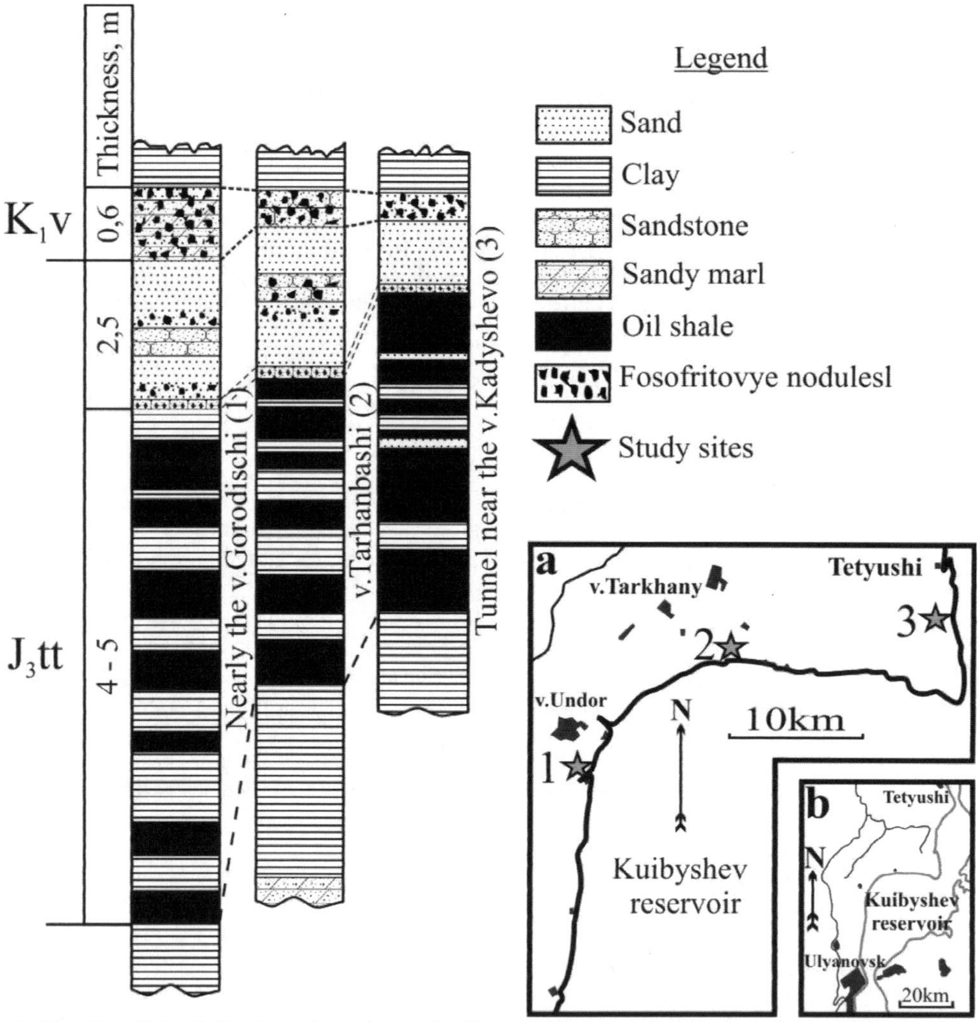

Figure 1. Combustible shale deposits schematic diagram in a sectional elevation within the interval of the village B.Tarhany - Undory according to [2].

2. Methodology

To assess the given fuel type adequacy for thermal energy production it is necessary to know its thermal characteristics and blend composition. The mentioned combined are to determine the combustible shale effectiveness as a fuel resource. In consideration of this the main research methods were X-ray, thermal and X-ray fluorescence analyses.

The X-ray analysis was carried out on the device SHIMADZU XRD-7000 in order to detail the samples composition. The shooting mode: range 3-60° by 2Θ, step 0.02 deg. / min. with the exposure lasted for 4 sec., copper radiation with the tube voltage 30kV. The standard powder compositions were used.

The thermal analysis was performed on the device STA 449 Jupiter F3 to determine the calorific value and phase transitions of the combustible shale structural components. The burning ranged from 30 to 1000 ° C, the heating step – 10 deg. / min. with constant air blowing.

The X-ray fluorescence analysis was performed with the spectrograph EDX-720P to determine the main rock chemical elements. The measurement accuracy was 0.001%.

3. Results and Discussion

Combustible shale in situ is laminated, cryptomerous, dark-grey and black rocks with inclusions of marine animals calcareous shells and pyrite nodules. In massif they are divided by vertical and horizontal cracks into separated blocks of various dimensions. The conducted physical and mechanical studies have shown that the rocks have similar properties to rocky ground species [3]. Most of them are characterized by relatively low porosity, moisture and density with relatively high strength parameters (Table 1).

Table 1. Combustible shale physical and mechanical properties.

Parameters	Letter value	Measurement units	Values		Average meaning
			from	to	
Natural humidity	W	%	1.5	6.0	3.2
Full possible humidity	Wo	%	6.0	12.0	8.6
Dry soil density	ρ	g/sm³	1.72	1.90	1.81
Porosity	n	%	6.0	10.6	8.3
Compressive strength	R_c	MPa	20.75	48.00	34.36
Tensile strength	R_t	MPa	0.83	1.92	1.37

According to the held X-ray analysis the combustible shale mineral part is characterized by multi-component composition (Fig. 2). The clay particles prevail – montmorillonite ($(Na,Ca)_{0.4}(Al,Mg,Fe)_{2-3}[Al,Si_3O_{10}](OH)_2{*}nH_2O$), muscovite ($KAl_2[Al,Si_3O_{10}](OH)_2$), chlorite ($(Mg,Fe)_3[Al,Si_3O_{10}](OH)_2$) and kaolinite ($Al_4[Si_4O_{10}](OH)_8$). Allogenic quartz grain (SiO_2) and albite ($Na[Al,Si_3O_8]$), and calcite ($CaCO_3$), represented by marine animals shell fragments, are present in smaller amounts. Out of authigenic substances minerals pyrite (FeS_2) is noted.

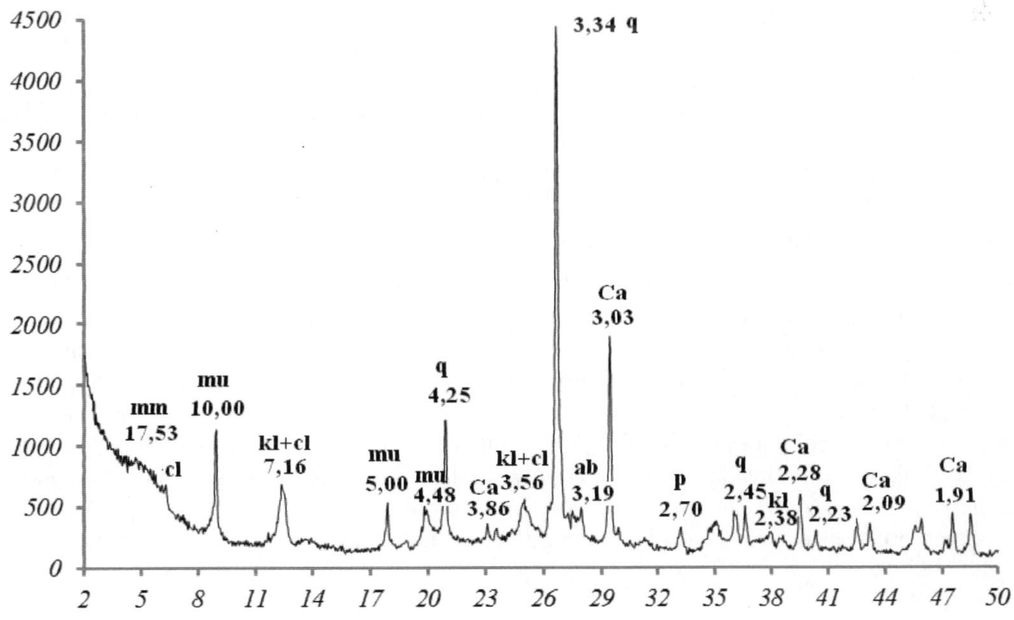

Figure 2. Distinctive combustible shale X-ray diffraction pattern: ***ca*** – calcite; ***q*** – quartz; ***mu*** - muscovite; ***cl*** – chlorite; ***kl*** – kaolinite; ***mm*** – illite-montmorillonite; ***p*** – pyrite; ***ab*** – albite.

According to the research data organic matter content in combustible shale varies from 8 to 40%. As for chemistry organic matters are represented with so-called kerogen – natural organic polymeric material with a high molecular weight. According to [2] yield of volatile substances when rocks are burnt is 39.2-77.25%. Out of these, CO_2 – to 20.7%; O_3 - to 3.7%; N_2 – to 14.9%; CO – to 0.8%; H_2 – to 10.6%; CH_4 – 49.6%; heavy hydrocarbons – up to 1.3%.

Minding furnace process technology it is necessary to know mineral and organic components behavior when combustible shale is burnt. Since the first will determine slagging properties of energy raw materials, the second – its calorific value and both of the mentioned will condition the carbon, nitrogen and sulfur with furnace gases release rate. In order to give predictive estimate of these parameters combustible shale thermal analysis was carried out. The obtained thermogram shows that various thermal conversions occur in rocks with sequential temperature increase of burning (Fig. 3). For data interpretation convenience there are two curves in the picture: a thermogravimetric one, showing the change in sample mass when heating, and a differential scanning calorimetry curve, showing the nature of absorption of heat (endo-effect) and liberation of heat (exo-effect).

The first sample mass loss occurs within the temperature interval from 60 to 150°C. Over this range in the differential scanning calorimetry curve two successive endo-effects are clearly manifested resulting from firstly capillary-condensation and then molecular (interlayer) bleeding from clay minerals.

The following changes are observed within the temperature interval from 150 to 500°C. Here in the differential scanning calorimetry curve two distinct exo-effects are observed accompanied by mass loss in the thermogravimetric curve. According to N.D. Topor [4], over this range organic matter thermal-oxidative breakdown occurs: within the interval from 150 to 400°C hydrocarbons light and middle fractions are oxidized, within the interval from 400 to 650°C – heavy fractions and kerogen. The area estimation of exo-effects equal to organic matter burnout, allows us to calculate the calorific value.

Further changes in mass within the temperature interval to 680°C are explained by hydroxyl water (OH^-) bleeding from the clay minerals structure. The implicit endo-effects in the differential scanning calorimetry curve is apparently due to imposed exo-effects caused by pyrite oxidation with the formation of hematite (α-Fe_2O_3). The mass loss within the temperature interval from 680 to 800°C, accompanied by an endo-effect, corresponds to thermal conversion of calcite CaO. The exo-effect in the differential scanning calorimetry curve with its maximum at the point 902°C passing without mass changes corresponds to recrystallization of amorphous clay minerals decomposition products, accompanied by a new mineral phase formation (spinellide, etc.) [5].

Ash combustion products of solid energy fuels have abrasive properties that cause severe wear of convective heating surfaces and other elements of flue gas path. To estimate the ash erosion parameters it is necessary to know the mineral composition of the combustible shale residual components. With this purpose in mind, a radiographic examination of ash residual components from the given rocks burning was conducted. In the obtained diffraction patterns strong diagnostic quartz, wollastonite ($CaSiO_3$), gehlenite ($Ca_2Al_2SiO_7$) and anhydrite ($CaSO_4$), less strong hematite lines are seen (Fig. 3). Taking into consideration a relatively high position of the background line concerning the zero mark, we can assume the amorphous substance presence in the ash constituent. All identified mineral phases are the thermal dissociation product of clay, iron and carbonate combustible shale components.

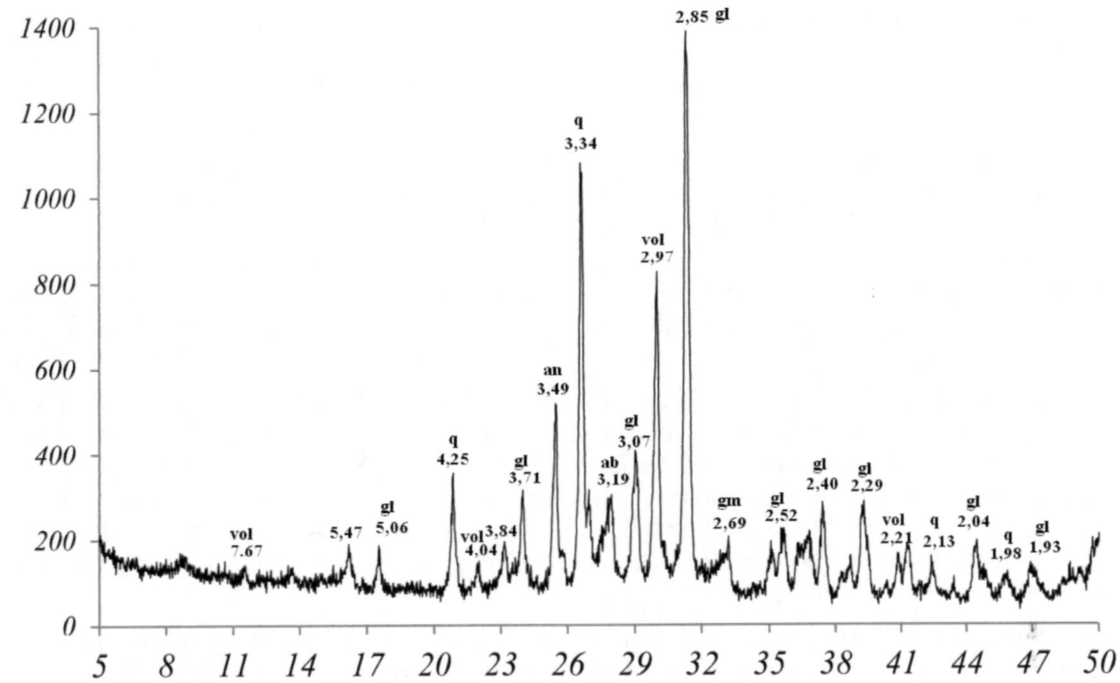

Figure 3. The X-ray diffraction pattern of the ash products of combustible shale: ***q*** – quartz; ***wol*** – wollastonite; ***gl*** – gehlenite; ***gm*** – hematite; ***an*** – anhydrite; ***ab*** – albite.

Considering that the properties of the ash combustion products of solid energy fuels are determined by the chemical composition, the X-ray fluorescence analysis of original combustible shale and its ash constituents, obtained after the rock burning at the temperature 1000°C, was held. The results, shown in Table 2, indicate that the noncombustible residual components are represented mainly by silica, calcium, alumina and iron oxides. The high sulfur dioxide content (SO_3) is noticeable.

These experiments allow us to give a predictive estimate of combustible shale engineering data as a power generation fuel. The physical-mechanical rocks properties analysis shows that their exploitation can be carried out both by digging and mining. At the same time the high strength properties of shale allow lump sampling technology to be used. This will enable tunnel kilns and box furnace to be used. If it is necessary to use pulverized pieces combustion the lumps can be powered at industry sites. The shale density and its low water saturation allow rocks to be powered at plant crushers without preliminary drying.

Table 2. The combustible shale and its ash combustion products according to the X-ray fluorescence analysis.

Sample	Main elements content in the oxide form, %								
	SiO_2	Al_2O_3	Fe_2O_3	TiO_2	CaO	MgO	Na_2O	K_2O	SO_3
Combustible shale	54.97	11.72	4.74	0.93	18.57	2.33	0.72	2.61	2.44
Ash constituent	55.28	10.51	5.12	0.98	17.44	2.33	0.81	2.75	3.82

Since the main solid fuels quality indexes are calorific value, moisture, ash and sulfur contents, we will consider the mentioned further.

The combustible shale calorific value ranges widely from 2.4 to 9.0 MJ/kg. In Tatarstan low-calorie differences with heating value 650-2134 cal. dominate, in the Ulyanovsk region rocks power-intensity increases and has 1400-3200 cal., and some samples have 5450 cal. The main rock combustible constituents are carbon and hydrogen. Oxygen and nitrogen are organic ballast.

The combustible constituent content per working fuel mass unit is determined by rock moisture and noncombustible mineral constituent amount. Additional moisture will take a certain amount of heat for its evaporation, reducing the combustible shale calorific value, which in fact determines the necessity of its accounting. Near surface rock positions will contribute to water saturation. However, the available porosity will cause a relatively small total water-absorbing capacity of rocks (W^p = 6.0-12.0%). Air-dried fuel, that is, with steady borrow-pit moisture content will contain 1.5-6.0% of gravity-capillary water.

The solid noncombustible residual component, formed after burning of the fuel, is an ash part of rock. According to the calculations the combustible shale ash content (A^d) varies from 60 to 80%, which agrees well with the datum given in the research [2]. By its composition shale ash is basic (CaO content exceeds 10%). The foundation modulus (hydrosilicate module) is M_o = 0.35, silicate (siliceous) module – M_c = 3.54, the quality factor – K = 0.54 . According to the set of indexes the Volga combustible shale ash is an active group, which means that it has the property to become solid spontaneously in the open air.

One of the negative aspects of solid noncombustible residual components is thermal effectiveness and performance degradation of boiler units, which is caused by slag deposits and abrasive ability of ash volatility.

The operating experience of the boiler units shows that the rate of deposits formation, in addition to furnace process variable, depends on the ash components chemical activity. Consequently the ash coking properties can be pre-assessed in terms of slagging (R), estimated on the basis of the noncombustible residual component chemical composition:

$$R = \frac{CaO+Mg+Fe_2O_3+Na_2O+K_2O}{SiO_2+Al_2O_3+TiO_2} \times Na_2O = 0.34 \qquad (1)$$

In the case of the Volga combustible shale burning, ash with low softening point due to the high alkali metals content in the rocks is expected to be obtained. Even at temperatures of 900 – 1000°C the slagging process will have an avalanche-like character. With the given X-ray analysis results it is possible to suggest that, in addition to an amorphous vitreous phase, the slag content will have a lot of spinel.

The ash abrasive properties include solid particles ability when colliding with pipe wall to cut off them microscopic metal layers. Due to this the pipe thickness gradually decreases, that can lead to breakthrough in the waste gas flue system. The ash wearing is primarily determined by the composition content of noncombustible residual constituents SiO_2 and Al_2O_3. Therefore, knowing the solid products chemical composition of combustible shale thermal conversion we can calculate the approximate ash abrasive coefficient [6]:

$$a = 0.045 \cdot (SiO_2 + Al_2O_3 - 44) \times 10^{-11} = 9.8 \times 10^{-10} m^2/N \qquad (2)$$

According to the obtained results ash got during the Volga combustible shale burning, will have relatively mild abrasive abilities.

The combustion technology for all fuels implies a high yield of gaseous components, some of which represent production problems. One of such components is sulfur. Therefore a special attention is given to the sulphur content of solid mineral fuel. According to the X-ray fluorescence analysis combustible shale contains up to 3.0% of total sulfur, most of which occur in pyrite. The considerable size of the pyrite nodules (1-6 cm) allows admitting that sulfur content is due to the presence of extracted sulfur pyrites from the rocks. During shale exploitation it is quite possible to use manual selection of bur units.

4. Conclusions

Summing up the obtained results, we can draw to the following conclusions:
1. The combustible shale in Tatarstan and the Ulyanovsk region has predominantly low calorific value (Q_b^d = 5-9 MJ/kg). In order to enhance its efficiency and to reduce cost it is possible to conduct rocks burning together with some other organic or organic mineral power generating fuels.
2. High ash content (A^d = 60-80%) that causes a high external ballast content in shale implies the appropriateness of using this fuel resource next to its exploitation site. The acceptable distance to a consumer will reduce unproductive transportation charges for large ash and moisture masses.
3. The performed fuel ash components characteristics, as well as the yield and volatiles composition allow us to specify the basic parameters for boiler units, designed for the Volga combustible shale burning.
4. The noncombustible residual components composition shows that shale ash can be used in manufacture of materials of construction.

Acknowledgments

This study was performed in the context of the Russian Government Program of Competitive Growth of Kazan Federal University.

References:

[1] V. G. Emeshev and M. S. Parovinchak, “Without imported energy”, Oil and gas vertical, vol. 17, pp. 63-65, 2005.
[2] V. A. Cherdyntsev and E. I. Tikhvin, "The geology of the Tatar ASSR and neighborhood within 109 sheet” (Moscow, Publishing house «GONTI»), p 204, 1939.
[3] GOST 25100-95 Soils. Classification. Russia
[4] N. D. Topor, L. P. Ogorodova and L. V. Mel'chakova, “Thermal analysis of minerals and inorganic compounds” , (Moscow, MSU Publishing house), p 190, 1987.
[5] V. P. Ivanov, B. K. Kasatov, T. N. Krasavina, E. L. Rozinova, “Thermal analysis of minerals and rocks”, (St. Petersburg, Publishing house «Nedra»), p 399, 1974.
[6] A. A. Abrutin and E. S. Karasina, “Thermal design of boilers (normative method)”, (St. Petersburg, Publishing house of NPO CKTI), p 256, 1998.

IOP Conf. Series: Earth and Environmental Science **155** (2018) 012012 doi:10.1088/1755-1315/155/1/012012

Drainage and impregnation capillary pressure curves calculated by the X-ray CT model of Berea sandstone using Lattice Boltzmann's method

T Zakirov, A Galeev and M Khramchenkov
Kazan Federal University, 18 Kremlevskaya Str., Kazan 420008, Russia

e-mail: tirzakirov@kpfu.ru

Abstract. The study deals with the features of the technique for simulating the capillary pressure curves of porous media on their X-ray microtomographic images. The results of a computational experiment on the immiscible displacement of an incompressible fluid by another in the pore space represented by a digital image of the Berea sandstone are presented. For the mathematical description of two-phase fluid flow we use Lattice Boltzmann Equation (LBM), and phenomena at the fluids interface are described by the color-gradient model.
Compared with laboratory studies, the evaluation of capillary pressure based on the results of a computational filtration experiment is a non-destructive method and has a number of advantages: the absence of labor for preparation of fluids and core; the possibility of modeling on the scale of very small core fragments (several mm), which is difficult to realize under experimental conditions; three-dimensional visualization of the dynamics of filling the pore space with a displacing fluid during drainage and impregnation; the possibility of carrying out multivariate calculations for specified parameters of multiphase flow (density and viscosity of fluids, surface tension, wetting contact angle).
A satisfactory agreement of the capillary pressure curves during drainage with experimental results was obtained. It is revealed that with the increase in the volume of the digital image, the relative deviation of the calculated and laboratory data decreases and for cubic digital cores larger than 1 mm it does not exceed 5%. The behavior of the non-wetting fluid flow during drainage is illustrated. It is shown that flow regimes under which computational and laboratory experiments are performed the distribution of the injected phase in directions different from the gradient of the hydrodynamic drop, including the opposite ones, is characteristic.
Experimentally confirmed regularities are obtained when carrying out calculations for drainage and imbibition at different values of interfacial tension. There is a close coincidence in the average diameters of permeable channels, estimated by capillary curves for different interfacial tension and pore network model. The differences do not exceed 15%.

1. Introduction

Capillary force plays a significant role at multiphase flows in a porous medium and an important characteristic in the study of hydrodynamic processes. The value of the capillary pressure is a function of the medium saturation by different phases and, in the first approximation, depends on the geometry of the pore space, the interfacial tension and the contact angle of wetting [1]. In laboratory conditions the most common methods for measuring capillary pressure curves are the mercury porosimetry [1], capillarimetry (or semipermeable membrane method) [2] and centrifugation [3]. The general drawbacks of these methods include such laborious stages as the preparation of fluids and the surface of pore channels, the long time of experiments on low permeable samples.
As an alternative to laboratory filtration experiments, the methods of numerical calculations on a digital microtomographic core images are being intensively developed in recent years. For computing resources optimization pore-network models (PNM) [4], in which the original digital image of pore space is associated with a model of spherical pores and cylindrical capillaries connecting them, are used. According to this model, with a given geometry of the pores and tubes capillary pressure is estimated by the Laplace formula. The increasing availability of high-speed computing technology and the compilers

improvement make it possible to use a computational experiment on immiscible displacement of one incompressible fluid by another in pore-space, which is represented by the original digital image. The features of this method are described in detail in Sections 3.
The correspondence problem between the filtration parameters, which are calculated during computer modeling on the digital cores models of the order of 1 mm in size, and the values obtained in laboratory experiments on standard samples with a size of several centimeters is the subject of discussion in works of many authors [4, 5]. According to one of the most common approaches [4], the question of rescaling from the micro to the laboratory level is solved by finding a representative elementary volume of the digital image, in which the characteristics of the microtomographic model reproduce the properties of the sample at a centimeter level. However, as shown in [5], this is not a universal approach in cases of a complex pore space structure, especially for carbonate reservoirs. Thus, using of simulation results on the digital cores for describing the properties of the specific reservoirs requires a particular methodological study, some theses of which are described in Ref [5].
The purpose of this paper is to verificate the proposed approach, which consists in quantitative and qualitative data comparison with known experimental results and regularities, as well as testing the appropriateness and adequacy of this technique for typically small core fragments used in microtomography, on the order a few millimeters. The sample of the study is a digital model of a natural sandstone with volume ~ 1 mm^3 which is representative in the laboratory scale for the coefficients of porosity and absolute permeability. This circumstance allows us to make a correct comparison of the calculated and experimental capillary pressure curves.
Compared with laboratory studies, capillary pressure estimation, which is based on the results of a computational filtration experiment, is a non-destructive method and has a number of advantages: the labor costs absence for preparation of fluids and core; the possibility of modeling on the scale of very small core fragments (several mm), which is difficult to realize under laboratory conditions; three-dimensional visualization the pore space filling with a displacing fluid during drainage and imbibition; the possibility of carrying out multivariate calculations for various parameters of multiphase flow (density and viscosity of fluids, surface tension, contact angle).

2. The sample of study.

The sample of the study is the Berea sandstone. This type of sandstone is standard for laboratory testing of oil and gas industry processes and has been used in various works, among which Tsuji T. et al. [6] Blunt M. et al. [7], Berg S. et al. [8]. According to the mineralogical studies given in [8], Berea sandstones consist of quartz, feldspar, calcite and clay minerals.
A digital model of pore-space from the Imperial Colledge London open library [http://www.imperial.ac.uk/earth-science/research/research-groups/perm/research/pore-scale-modelling/micro-ct-images- and-networks / berea-sandstone /] is represented by a cubic binarized three-dimensional microtomographic image with size of 400 voxels and a resolution of 3.2 μm. Images of Berea sandstone in 3D and 2D projections are shown in figure 1. The porosity coefficient of this fragment is 19.6%. The absolute permeability tensor is practically isotropic, since the fraction of its isotropic component K_{iso} = 0.670 μm^2 [9, 10] is more than 92%. The cube fragment of Berea sandstone with a size of 1 mm is representative, and this, in fact, is slightly smaller than the volume of the image we use, whereas experiments for the capillary pressure curves measuring were carried out on a 160 mm^3 sample (10 mm × 4 mm × 4 mm).

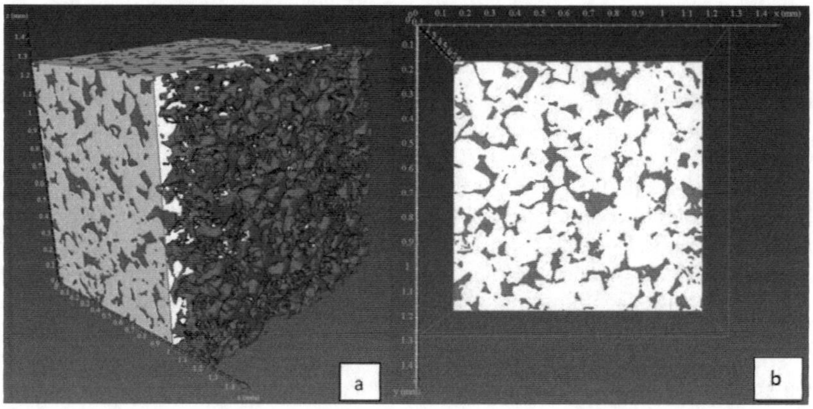

Figure 1. The binarized digital model of Berea sandstone with the size of 400 × 400 × 400 voxels and resolution 3.2 μm: a – 3D model; b – slice in XY plane ; white - skeleton, blue - pore space

3. Materials and methods.

3.1. Method for capillary pressure curves measuring during drainage and imbibition.

In this paper for the capillary pressure curves determination we carry out a computational experiment on the immiscible displacement of one incompressible fluid by another. At the initial time, the pore space is completely filled with the wetting phase. The computational experiment consists of several stages. The first consists in injecting a nonwetting phase into the sample and, consequently, displacing the wetting fluid from it (drainage). The injected nonwetting fluid is fed to the leftmost side (input boundary), which is perpendicular to the OX axis, and the selection of two liquids occurs through the rightmost section (the exit boundary), also perpendicular to the OX axis. The other four outer faces of the computational domain are impermeable to fluids. We carry out the experiment at various pressure differences between the inlet and outlet sections. The minimum pressure drop at which the displacement process begins is a threshold. At the finishing of each mode, i.e. when a flow rate of the wetting phase through the outlet boundary is zero, we make a stepwise increasing in pressure at the inlet section, and the drainage process continues. The drainage process finishes when the wetting phase displacement with a further increase in pressure terminates.

At the next stage, the pressure at the input boundary is reduced. All cells in the outlet section that are related to the pores being filled with the wetting phase. During this experiment, the wetting phase displaces the nonwetting through the inlet region of the sample, i. e. we simulate the imbibition process. When the flow rate of the nonwetting phase through the inlet section is zero, the pressure on it decreases in a stepwise manner, after which the imbibition continues. In the latter stage of imbibition, the pressure difference between the input and output boundaries are equal. After the calculations under the given conditions, the experiment is completed.

The measurement of sample saturation by wetting phase at each drainage and impregnation regime is carried out under the condition when wetting and nonwetting phases are not displaced, respectively, and the resulting pressure drop is therefore capillary.

3.2. Lattice Boltzmann method.

The paper considers a two-phase isothermal flow of immiscible incompressible liquids in porous medium. The Lattice Boltzmann method or Lattice Boltzmann Method equations (abbreviated LBE) are used to describe the flow of liquids. This mathematical model is described in many works [11, 12], so we will not give a detailed mathematical description of it, but dwell only on the basic assumptions.

In the framework of LBE, the flow of a medium is considered as dynamics of a particles ensemble with a given finite number of possible velocities. The flow domain is represented by grid with a square or cubic cell shape. The set of these form a lattice. During a time step Δt particles without interaction with each other can make one act of transition between adjacent lattice sites. One-particle distribution functions $f(\mathrm{r}, \boldsymbol{u}, \mathrm{t})$ are used to describe the state of the system at each grid node. This function shows the part of particles at time t located n the vicinity of point $r(x,y,z)$ with coordinates from x to $x+\Delta x$, from y to $y+\Delta y$, from z to $z+\Delta z$ and with velocity range from $\mathbf{u}(u_x, u_y, u_z)$ to $\mathbf{u}(u_x+\Delta u_x, u_y+\Delta u_y, u_z+\Delta u_z)$.

For a two-dimensional flow domain, for example, a discrete set of velocities D2Q9 is given as follows: $e_1 = c\cdot(0,0)$; $e_2 = c\cdot(1,0)$; $e_3 = c\cdot(0,1)$; $e_4 = c\cdot(-1,0)$; $e_5 = c\cdot(0,-1)$; $e_6 = c\cdot(1,1)$; $e_7 = c\cdot(-1,1)$; $e_8 = c\cdot(-1,-1)$; $e_9 = c\cdot(1,-1)$, where $c = \dfrac{\Delta l}{\Delta t}$ – the lattice speed, Δl – the grid spacing. Each velocity vector from a given set e_i (i=1..9) corresponds to the distribution function $f_i(r,t)$, depending only from t and r. In a case of 3D we use D3Q19 model.

The dynamics of the particles ensemble of each fluid is described in several stages. The first is called streaming step. At this stage, the particles move to neighboring nodes during time Δt, and the direction of the motion speed does not change. The second stage deals with the collision process of particles, as a result of which the distribution function tends to the equilibrium state. The third stage describes the interface interaction between two fluids and between fluids and the solid phase. The distribution function evolution for each fluid in time and space is described by equation (1):

$$f_i^k(r+e_i\Delta t, t+\Delta t) = f_i^k(r,t) + (\Omega_i^k(r,t))^1 + (\Omega_i^k(r,t))^2, \quad (1)$$

where $k = o, w$ – are oil and water.

$(\Omega_i^k)^1$ is a collision operator. There are two main models, which describe collision of particles – Single Relaxation Time (SRT) [11-13] and Multi Relaxation Time (MRT) [13]. The results obtained with MRT model use have a more exact match when comparing with known analytical solutions [13] (for example, Couette flow) than SRT model. So, that is why in our work we use MRT approach.

The relaxation coefficient τ^k is the determining parameter in the SRT and MRT models and controls kinematic viscosity according to the equation (2):

$$\mu^k = \left(\frac{2\tau^k - 1}{6}\right)\frac{\Delta l^2}{\Delta t} \quad (2)$$

In contrast to the classical equations of the fluid and gas mechanics (for example, the Navier-Stokes equations), where the solution is sought in the "pressure-velocity" variables, the LBE are solved in the "density-velocity" variables. The macroscopic density and velocity in the cell of domain are calculated using (3) and (4) formulas, respectively:

$$\rho^k(r,t) = \sum_{i=1}^{9} f_i^k(r,t) \quad (3)$$

$$u^k(r,t) = \frac{1}{\rho^k}\sum_{i=1}^{9} e_i f_i^k(r,t) \quad , \quad (4)$$

The pressure p^k in LBE, which is produced by each fluid, is related to its density by the following relation: $p^k = \dfrac{\rho^k c^2}{3}$ [11, 12].

The phenomena at the interface of fluids are described using the color-gradient method [14]. It consists of several types:

1) calculation of the color field gradient $\mathbf{g}$, which components are calculated by the formula:

$$g(r,t) = \sum_{i=1}^{9} e_i (f_i^o(r+e_i\Delta t, t) - f_i^w(r+e_i\Delta t, t)). \quad (5)$$

Traditionally, one fluid is red, and the second is blue;

2) description of the surface tension effects on the interface between fluids:

$$(\Omega_i^k)^2 = \frac{A}{2}|\mathbf{g}|(2\cdot\cos^2(\alpha_i) - 1), \quad (6)$$

where parameter A controls surface tension σ, α_i is an angle between $\mathbf{g}$ and e_i direction.

3) «recoloring» step – modification of f_i^k after solving equation (1):

$$(f_i^o)^* = \frac{\rho^o}{\rho} f_i + \beta\frac{\rho^o\cdot\rho^w}{\rho} f_i^{eq}\cdot\cos(\alpha_i) \quad (7)$$

$$(f_i^w)^* = \frac{\rho^w}{\rho} f_i - \beta\frac{\rho^o\cdot\rho^w}{\rho} f_i^{eq}\cdot\cos(\alpha_i) \quad (8)$$

,

where $\rho = \rho^o + \rho^w$, $f_i = f_i^o + f_i^w$, f_i^{eq} is an equilibrium function distribution [11], calculated at a density ρ and zero velocity. The parameter β influences on thickness of the fluid interface. In this paper, its value is 0.8 and can not exceed 1.

On the internal solid and external impermeable boundaries of the flow domain we use "bounce back" conditions [11, 12]. At the input and output boundaries, the fluid pressure is known and the velocity components normal to the boundary is equal to zero. In LBE, such conditions are given by Zow and He equations [15].

The densities of the wetting and nonwetting phases in all computational experiments are 1000 kg / m^3, and the viscosity is 2 mPa·s and 1 mPa·s, respectively. According to [8], the contact wetting angle is 0^0. This model assumes a constant temperature, chemical composition, no deformation of the skeleton and the density of liquids in the used pressure range.

4. Results and discussion

To verify the method proposed in Sections 2 and 3, we perform a comparison of the capillary drainage pressure curves, the calculation of which was carried out on the digital core samples with dimensions of 100, 200, 250 and 300 voxels, with the laboratory experiment results (figure 2). The digital core tested in our study was equal to a sample for which laboratory measurements of porosity, absolute permeability, and the capillary pressure curves by mercury porosimetry were made in Ref. [8]. The surface tension in the calculations and experiment was 28.8 mN / m. For each curve, the average of the relative deviation absolute value was calculated using the following formula: $\Delta^{\alpha} = \frac{100}{N} \cdot \sum_{i=1}^{N} \left| \frac{S_i^{\text{model}} - S_i^{\text{experiment}}}{S_i^{\text{model}}} \right|$, in which N is the number of pressure measurements (in our case N = 11); S_i^{model} and $S_i^{\text{experiment}}$ – the wetting phase saturation at i-th capillary pressure, measured in the computational and laboratory experiments, respectively; α indicates the size of the digital core. According to the results shown in figure 2, Δ^{100} = 10.2 %, Δ^{200} = 8.5 %, Δ^{250} = 5.0 %, Δ^{300} = 3.3 %. The fact that Δ for model with size of 250 voxels and above does not exceed 5%, the cubic digital cores of this sample, taken with a resolution of 3.2 μm and sizes of more than 1 mm, can be considered representative on a laboratory scale for measuring the capillary pressure of Berea sandstones. Note that the size of the representative volume is a specific value for each type of pore-space and can even exceed the dimensions of the digital model.

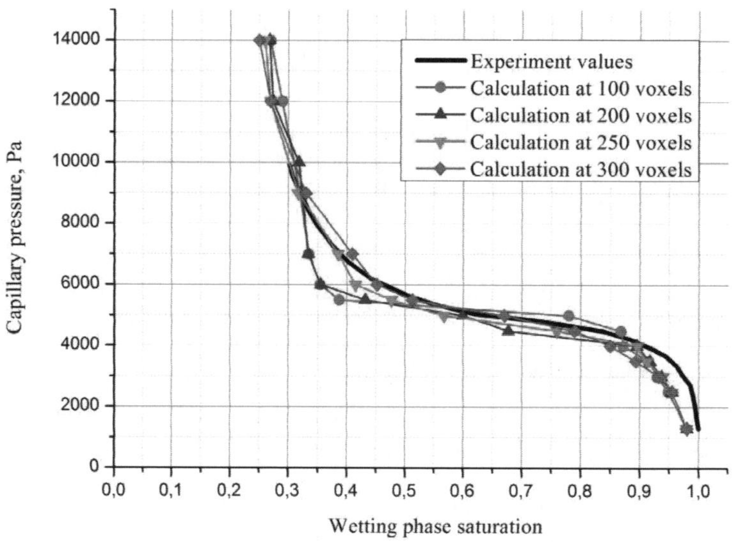

Figure 2. Comparison of the capillary pressure curves measured experimentally and calculated on the digital cores cubic fragments with sizes of 100, 200, 250 and 300 voxels

Figure 3 illustrates the dynamics of pore space filling with a non-wetting phase during drainage for the 100-voxel fragment. In figure 3b, 3c and 3d, the yellow arrow indicates areas in which the flow directions of the non-wetting phase differ from the created pressure gradient (i.e. upward) including the opposite ones (figure 3c and figure 3d). Analyzing the flow pattern in figure 3a and figure 3b, it can be noted that the interface coordinate, which is closest to the exit cross section, has not changed, and the displacement of the wetting phase occurs approximately perpendicular to the hydrodynamic pressure gradient. Such effects are a distinguishing feature of the regimes in which "capillary" fingers are formed and arise in cases when capillary number is less, than –3 [6].

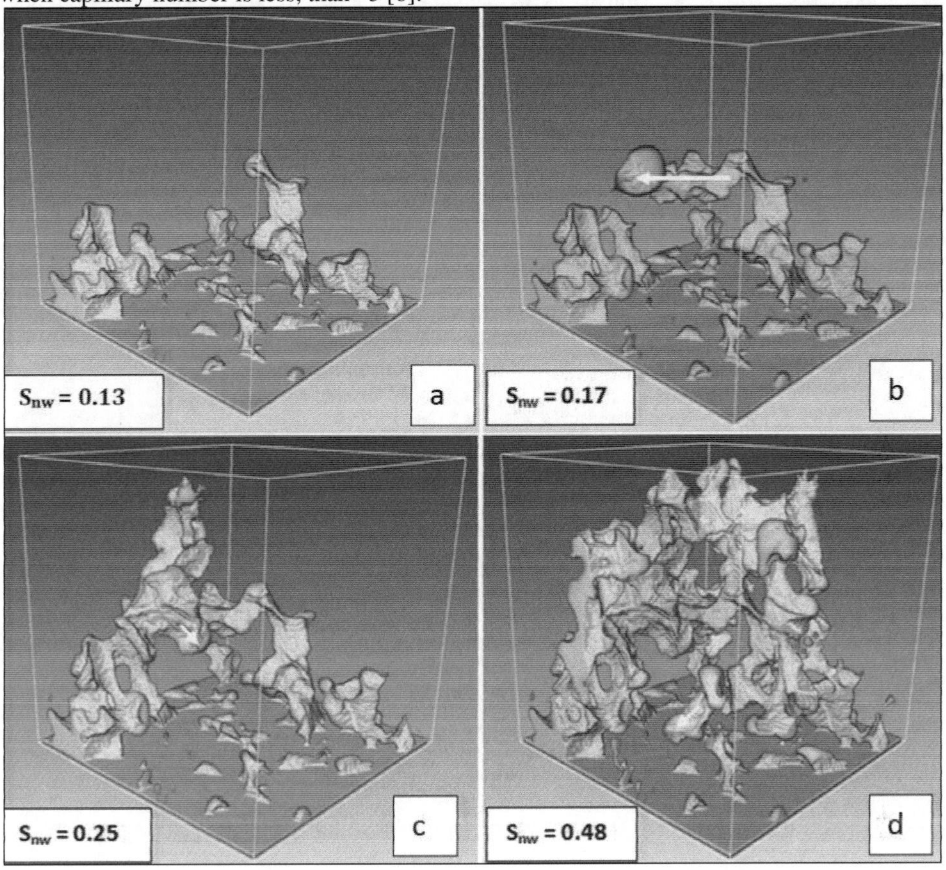

Figure 3. Visualization of non-wetting phase flow during drainage. S_{nw} - the volume fraction of the non-wetting phase in the pore space: a – 0.13; b – 0.17; c – 0.25; d – 0.48

Further, we investigate the trend of increasing capillary pressure in the series of computational experiments with different values of the interfacial tension σ = 7.7 mN/m, 14.4 mN/m, 28.8 mN/m. The calculations of drainage and imbibition curves were performed on a digital core of a cubic shape with sizes of 250 voxels. The results (figure 4 and Table 1) show that the drainage and imbibition curves measured at the same interfacial tension are not equivalent to each other, i.e., the hysteresis is manifested. Reduction of interfacial tension causes a decrease in capillary pressure, resulting in both the drainage and the imbibition curves are shifted to smaller values. The values of the residual saturation of a wetting phase (column 3 in Table 1) and the threshold drainage capillary pressure (column 2 in Table 1) are reduced by decreasing the interfacial tension, and the residual saturation with a nonwetting phase (column 4 in Table 1) increases with impregnation. Figure 5 shows the distribution of the residual nonwetting phase in the sample when impregnated for 14.4 mN/m. As can be seen from figure 5, the liquid is mostly presented the closed globules enclosed in the pore channels, some of them are shown in a larger size on the right-hand side of the figure 5. However, the results obtained by the above-mentioned method, are in agreement with experimental results [16].

We made comparison of the pore channels average diameters, estimated on the drainage results using the Laplace formula ($P_C = \frac{2\sigma}{R}$) and in direct measurement by PNM [6]. The results are demonstrated in columns 5 and 6 of Table 1 (the relative deviation from the average pore diameter by PNM is indicated in parentheses in column 5). Since the drainage curve at the interfacial tension 28.8 mN / m was successfully compared with the experimental (Δ^{250} = 5.0%), and the calculated diameters for different interfacial tensions are close (column 5), we can conclude that Laplace's formula estimates the pore size distribution with sufficient accuracy.

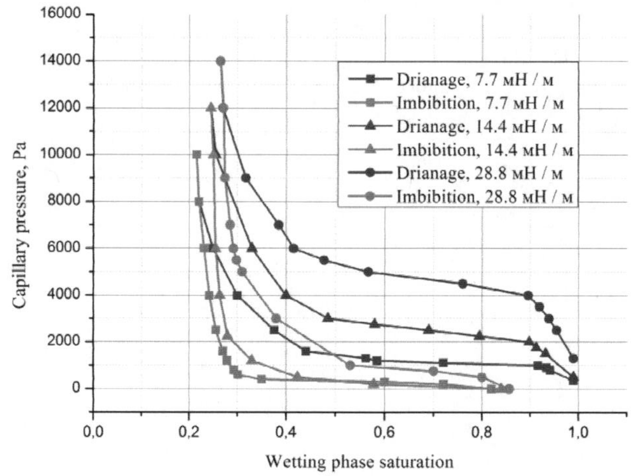

Figure 4. Capillary drainage and impregnation pressure curves at various interfacial tension

Table 1. Characteristics of capillary pressure curves at different interfacial tensions

Interfacial tension, mN/m	Threshold pressure, Pa	Residual saturation of a wetting phase, relative units	Residual saturation of a nonwetting phase, relative units	Average pore diameter in LBM (the relative deviation from PNM), µm	Average pore diameter in PNM [6], µm
28.8	1300	0.265	0.143	18.1 (–9.5 %)	20
14.4	500	0.245	0.16	17.4 (–13 %)	
7.7	350	0.215	0.18	18.1(–9.5 %)	

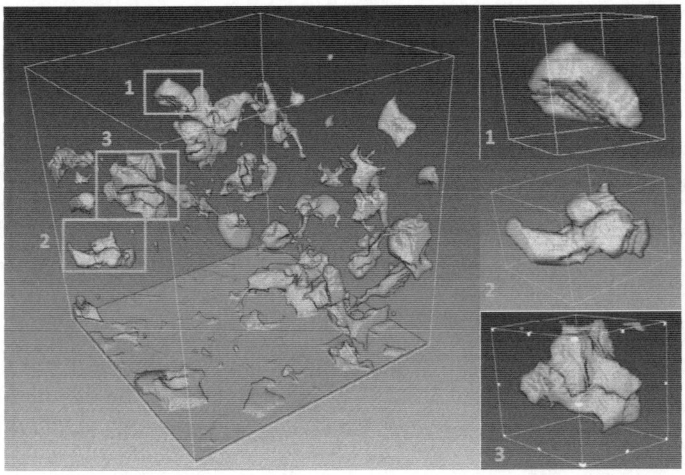

Figure 5. Distribution of residual non-wetting phase at 14.4 mN / m.
Some of isolated globules are shown on the right-hand side in larger scale

5. Conclusions

In the work, testing and approbation of the capillary pressure curves measuring technique by microtomographic digital images was carried out. We obtained a satisfactory agreement with the experimental data. It was revealed that with the digital image volume increasing, the discrepancy between the calculated and experimental data decreases.

The pore channels system of Berea sandstone in a cubic digital model larger than 1 mm is representative in laboratory scales for the estimation of capillary pressure values.

The pictures of the non-wetting liquid flow during drainage are illustrated. It was shown that for the flow regimes under which the computational and laboratory experiments are carried out, the directions of nonwetting phase flow differ from the gradient of the hydrodynamic pressure drop, including the opposite ones.

During drainage and imbibition calculations at different values of interfacial tension, we obtained regularities, which are confirmed experimentally. Comparison of average pore diameters estimated by capillary curves for different interfacial tension and pore-network model is satisfactory, the discrepancy does not exceed 15%.

Acknowledgments

The work is performed according to the Russian Government Program of Competitive Growth of Kazan Federal University and funded by RFBR according to the research project mol_a №"16-35-00155"

References

[1] Gimatudinov Sh 1971 *Physics of the oil and gas reservoir* (Moscow: Nedra)
[2] Masket M 2004 *Physical foundations of oil production technology* (Moscow: Izhevsk)
[3] Kuznetsov A M et al 2010 Neftyanoe khozyaistvo - Oil Industry **01** 49
[4] Blunt M J 2013 Pore Advances in Water Resources **51** 197
[5] Dvorkin J 2011 Geophysics **76** E141
[6] Tsuji T 2016 Advances in Water Recources **95** 3
[7] Mostaghimi P 2013 Mathematical Geosciences **45** 103
[8] Leu L 2014 Transport in Porous Media **105** 451
[9] Zakirov T R 2016 Current science **110** 2142
[10] Zakirov T R 2015 Neftyanoe khozyaistvo - Oil Industry **8** 36
[11] Pan C 2006 Computers and Fluids **35** 898
[12] Chen S 1998 Annual Review of Fluid Mechanics **30** 329
[13] Aslan E 2014 International Journal of Materials, Mechanics and Manufacturing **2** 317
[14] Reis T 2007 Journal of Physics A: Mathematical and Theoretical **40** 4033
[15] Zou Q 1997 Physics of Fluids **9** 1591

[16] Demond A H 1991 Water Resourses Research 10.1029/90WR02408

IOP Conf. Series: Earth and Environmental Science **155** (2018) 012013 doi:10.1088/1755-1315/155/1/012013

The free radical chain mechanism of the initial stages of crude oil oxidation in term of SARA fractions

A Ushakova[1], D Emelyanov[1], V Zatsepin[2], M Varfolomeev[1]

[1] Department of Physical Chemistry, Kazan Federal University, Kremlevskaya str. 18, 420008 Kazan, Russia
[2] Institute of Geology and Petroleum Technologies, Kazan Federal University, Kremlevskaya str. 18, 420008 Kazan, Russia

E-mail: paravoz-s@yandex.ru

Abstract.
The formation and decomposition of hydro-peroxides are the key stages of combustion. These stages strongly depend on the several factors accelerating or slowing this process. The aim of this work is to estimate experimentally which oil components act as inhibitors of initial stages of oxidation and which accelerate the process. The next aim is to explore the process of adsorption of oil components on the grain of rock, which turned to be also a key process in the low temperature oxidation.
The work includes experimental part where differential scanning calorimeter (PDSC) experiments with pure saturates, mixtures of saturates and aromatic oil fractions and mixtures of saturates, aromatic fractions and rock samples are considered. Effects of inhibition and acceleration of the initial oxidation stages are explored.

1. Introduction
It is known that predicting the performance of enhanced oil recovery by air injection has remained a challenge. It occurs mainly because of the lack of accurate models of crude oil oxidation.
In the recent works crude oil oxidation for the application of in-situ combustion was considered from the positions of the free-radical reactions, including radicals and short living hydroperoxides [1, 2]. Such consideration enables us to drive out some parameters like induction time that experience a huge impact from inhibition and acceleration effects [2].
The chemical structure of hydrocarbons effects dramatically on the hydroperoxides' activity and free radicals formation. Chain mechanism of oxidation is impossible to consider without dividing oil components into some groups according to their structure. Because of the complexity of cure oil oxidation a great number of chemical models were introduced. In recent years the investigators focused on SARA fraction model for oil oxidation [3-5]. The differences in the oxidation behavior of each SARA fraction for the particular oil and its impact on the overall oxidation of the oil have been studied in the work [3]. In the work [4] it was suggested to use RTO experimental results for each SARA description and reaction constructing. The attempt to introduce hydroperoxides into the oxidation model of SARA fractions was done in the work [5]. But the main drawback of such complicated model was

the impossibility of identifying kinetic parameters for such reactions using combustion tube or RTO experiments.

As it was shown experimentally [6] the main fuel for low temperature oxidation is pure saturates. Other components has much higher temperature of combustion and slow the oxidation reaction of saturates. The similar results were obtained in [7] using PDSC. This work also has shown the different effect of rock sample upon the SARA fractions' oxidation.

In the previous work [2] we have considered factors which are significant for self ignition process using model of SARA fractions. The initial stages of oxidation were considered from a position of radical chain process. We estimated the most valuable reactions by making a correlation among the oxidation stages and heat release obtained from experiment of oxidation with differential scanning calorimeter. The sequence of chemical reactions including free radicals, hydro-peroxides and inhibitors influence were considered.

The aim of this article is to clarify what kind of inhibitors act upon hydrocarbon oxidation process, which contain in the crude oil. We also considered the influence of crashed core samples on the oxidation in order to clarify if there are some kinds of oxidation catalysts in core material, or the core surface act upon the oxidation.

The differential scanning calorimetric technique was used for the oxidation experiments with oil saturates, oil aromatic components, their mixture and crashed core samples.

2. The main idea of the laboratory experiments.

It is well known that crude oil is multycomponent system, and oxidation of different oil fractions happens with different reaction rates and initial temperatures. There are two distinct areas for oxidation of any crude oil. The heat output increase in two temperature ranges 150-300 ^{0}C (low temperature oxidation LTO) and 350-500 ^{0}C (high temperature oxidation HTO). These areas one can easily observe using PDSC.

The catalytic or inhibitor effects come out mainly in the initial stages of oxidation. The inhibitors dramatically effect on the dependence of the initial heat appearance on the temperature. And this ability will be explored in the paper.

The papers mentioned above [2, 6] have shown that in the LTO region the saturate oil components exhibit oxidation. For light oils saturates make the main contribution into the heat release, the oxidation process for light oils is similar to saturates behavior. That is why the inhibition and catalytic effects were explored using pure saturates (paraffinic compounds) with different additives.

The experiments were designed in order to determine the influence of crashed rock on the oxidation of hydrocarbons as a catalyst or as an adsorbent of oxidized compounds. The possible catalytic effect of the rock on the oxidation and the inhibitory effect of some oil components suppressed by the rock were investigated separately.

The experiments were held with PDSC technique under 2,5 – 7 MPa pressure. The PDSC method allows making accurate measurements of heat release in constant flow of air with constant heating rate. The sample of 0,3 – 0,6 mg was hydrocarbone liquid or liquid with crashed rock, mixed in the ratio 1/15. The rock was taken from the core sample of the oilfield, which oil components were examined.

The oxidation of pure synthetic saturates C_{24}-C_{54} with linear structure was explored. Then tetracosane $C_{24}H_{50}$ oxidation curve was taken as a reference one, then C_{24} was mixed with oil aromatic components in concentrations 1/9 and 3/7 in order to observe inhibition process. Then mixtures of $C_{24}H_{50}$ with crashed core and mixtures of $C_{24}H_{50}$ with aromatics and with crashed core were explored.

The influence of different additives on the kinetic parameters was explored and the influence on the initial stages of oxidation was explored as well. The induction period – time period from the beginning of oxidation till the first measured heat release. Its value depends on the oxidation conditions and strongly depends on the temperature of the first heat release according to the work [7] in the following way:

$$\tau_{ind} \sim \exp\left(-\frac{1}{T}\right). \qquad (1)$$

In coordinates $Ln(\tau_{ind})$ vs $1/T$ it is a straight line. This dependence can be used for estimation the inhibition and catalytic effects in the initial stages.

The other valuable feature of oxidation process is the curvature of the kinetic curve in the area close to zero. The development of oxidation process in time after completion of the induction period shows an exponential dependence:

$$dq/dt \sim \exp(\varphi t), \qquad (2)$$

where coefficient φ is proportional to the quantity of active free radicals in the oxidizing system [8] at the given temperature. This coefficient as well as induction period is a significant characteristic of the activity of initial oxidation stages. Both these factors are strongly affected by inhibitors. The induction period increases with a decrease in the number of active radicals, the rate of the exponential time-dependence decreases if the activity of the radicals becomes insufficient. These two criteria are applied in this work.

3. The experiments of synthetic saturates oxidation

The oxidation of saturates was carried out by PDSC. The pressure effect and molecular weight effect on the oxidation process and the kinetic parameters determined using PDSC. Heat output of saturates' oxidation process starts at similar temperatures for C_{24}, C_{20}, C_{30}, C_{40}, C_{50}. Its value depends on the heating rate, but roughly lays in the range 195-219 ^{0}C. A slight difference has C_{40}, C_{50} in the direction of decreasing the initial temperature. Values of initial temperature at different heating rates are given in table 1.

Table 1. Onset temperature of combustion curves under different heating rates (5, 10, 15 20 ^{0}C/min) DSC of high-pressure analysis data for different alkanes.

heating rate (C^0/ min)	$C_{20}H_{42}$	$C_{24}H_{50}$	$C_{30}H_{62}$	$C_{40}H_{82}$	$C_{54}H_{110}$
5	200,40	200,46	200,95	195,43	196,60
10	209,68	209,07	209,80	205,15	205,49
15	215,47	214,80	214,97	211,29	209,77
20	219,22	218,91	218,93	214,71	214,72

There also were determined integral heat release and activation energies, pre-exponents and all the kinetic parameter as well for each saturate oxidation. These values characterize the kinetic of the whole process. But the values are not decisive and do not reflect the kinetics of the initial stages of oxidation. That's why then we use only characteristics for initial stages and not the kinetic parameters for the whole process.

The induction period depending on temperature was determined for a set of kinetic curves at different heating rates. The coefficient φ in the exponential dependence of the kinetic curve on

time was also identified (φ_{C24} = 0,041±0,014). These parameters are used later for comparison.
Different alkanes have similar kinetic curves of oxidation, quite close values of initial and peak temperature, activation energies. Their characteristics are also similar to the kinetic curves for saturates extracted from crude oils [6]. Therefore, to further study the inhibitory and catalytic effects on used tetracosane $C_{24}H_{50}$

4. The oxidation experiments of tetracosane mixed with aromatic compounds

Aromatic compounds were separated from the studied oil using SARA analysis. Tetracosane is a solid at room temperature, therefore, for the preparation of mixtures of aromatics with tetracosane they were both previously heated. Were investigated the oxidation curves for a mixture of tetracosane/aromatics 9/1 and 7/3 at three pressures of 2,5, 5 and 7,5 MPa (figure 1) and at different heating rates 5, 10, 15, 20 ^{0}C/min.

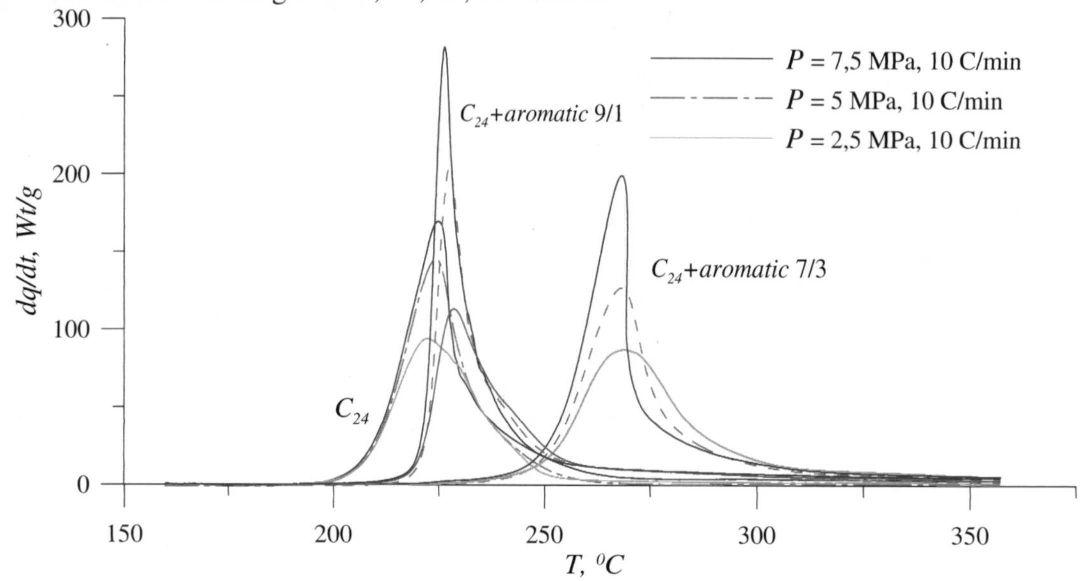

Figure 1. $C_{24}H_{50}$ and mixtures with Aromatics in 9 to 1 and 7 to 3 ratio combustion curves under 2.5, 5, 7.5 MPa.

It is well seen that the addition of aromatics inhibit the initial stages of tetracosane oxidation. The beginning of the oxidation is shifted to higher temperature. Oxidation temperature shifts increases from 200 to 230 ^{0}C for a ratio of 9/1 and up to 240 ^{0}C for a ratio of 7/3. Values of initial temperature at different heating rates for the mixtures are given in table 2. The initial stages of oxidation go slower. The first heat rise is smaller for mixtures the φ factor reduces from φ_{C24}= 0,041±0,014 to $\varphi_{C24+Arom}$= 0,026±0,011 (figure 4).
The effect of pressure is to increase the oxygen concentration and, accordingly, to increase the thermal effect of the reaction that has almost no effect on the beginning of the oxidation process and the heat rise of the initial stages (φ factor does not depend on the pressure). From this we can conclude that for saturates and saturates mixed with aromatics the appearance of hydroperoxides and free radicals occurs in the liquid phase during the initial stages of oxidation. The gas phase, entrained by the air steam in the flow mode does not affect the oxidation process.

5. The oxidation experiments of tetracosane mixed with crushed rock and tetracosane mixed with aromatics and crushed rock.

Further the oxidation of tetracosane and mixtures of tetracosane with the aromatics on crushed core material were investigated. We have taken the carbonate rock with the hydrocarbons in a ratio of 15/1, which corresponds to the average saturation of the considered oilfield.

To determine the effect of the rock were conducted experiments of C_{24} oxidation mixed with the crushed rock. Tetracosane was mixed at temperature 50-70 ^{0}C. The initial temperature of oxidation has not changed, but the pattern of the kinetic curve was changed to a more significant thermal effect at the initial stages of oxidation.

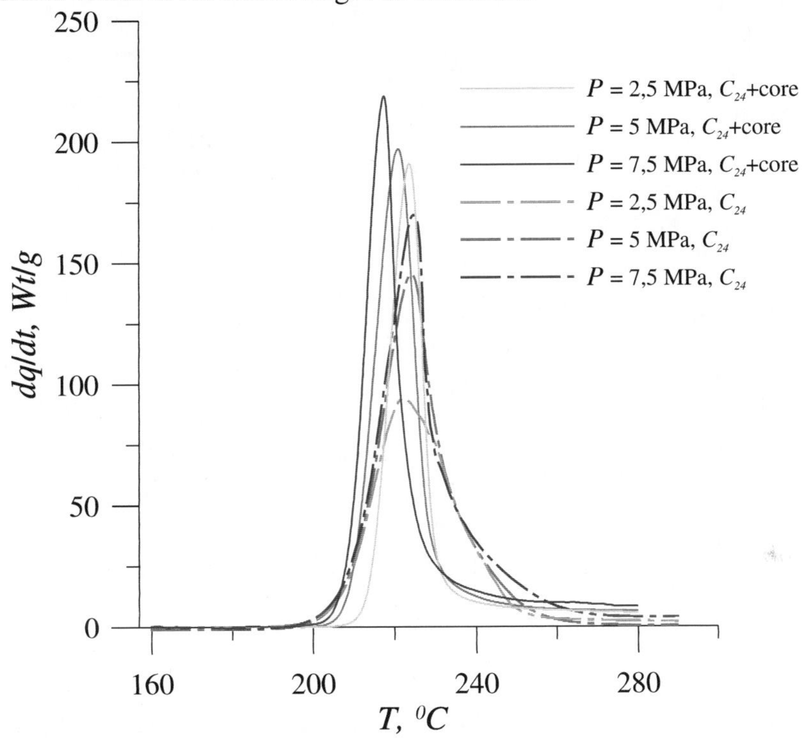

Figure 2 $C_{24}H_{50}$ and $C_{24}H_{50}$ mixed with crushed core combustion curves under 2.5, 5, 7.5 MPa.

One can observe the acceleration of oxidation at the initial stage with core material in comparison with pure tetracosane (figure 2) φ_{C24}= 0,041±0,014, and $\varphi_{C24+core}$= 0,158±0,009. The initial temperature of the heat output and accordingly, the induction time has not changed. But one can see a dependence of the induction period of the pressure for tetracosane mixed with core. The higher the pressure, the smaller induction time we observe (table 2).

The next step was to examine mixture of aromatic compounds and tetracosane with rock material. In the presence of crushed rock the aromatic compounds with tetracosane have a much lower initial temperature of heat release, it reduces by twenty degrees from 250 ^{0}C to 230 ^{0}C (table 2, figure 3).

Table 2. Onset temperature of combustion curves under different pressures (2.5, 5, 7.5 MPa) DSC of high-pressure analysis data for tetrakozane, tetrakozane with aromatics and with crushed core.

heating rate (C0/ min)	$C_{20}H_{42}$	$C_{20}H_{42}$ + core	$C_{24}H_{50}$ + aromatic 7/3	$C_{24}H_{50}$ + aromatic 7/3 +core
25	207,12	215,28	249,91	230,68
50	209,07	210,22	244,94	228,43
75	208,77	208,02	252,52	226,68

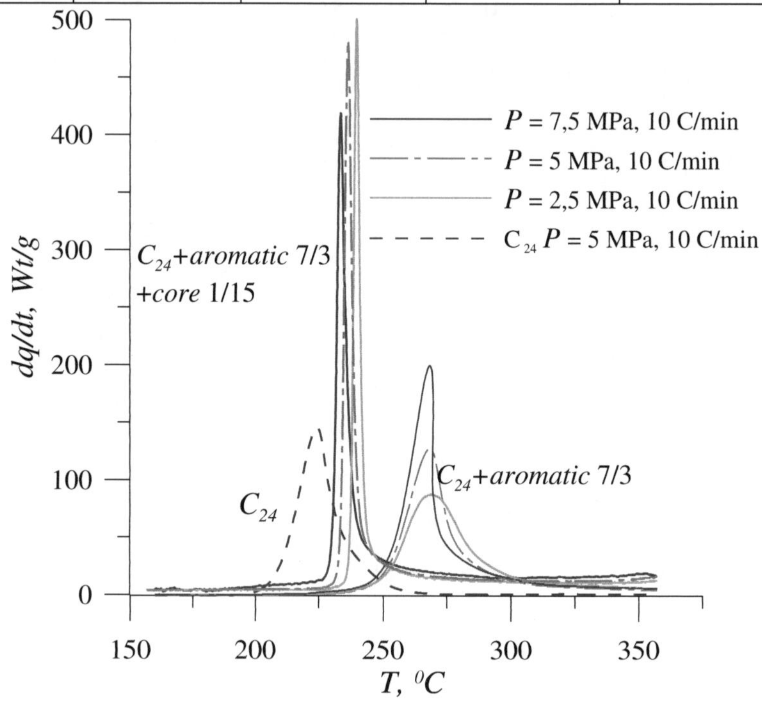

Figure 3. Pure $C_{24}H_{50}$ and $C_{24}H_{50}$ mixed with aromatics, $C_{24}H_{50}$ mixed with aromatics and crushed core, combustion curves under 2.5, 5, 7.5 MPa.

The induction time for mixtures on the rock also decreases. Such effect was not observed in earlier experiments with pure tetracosane (figure 2). Thus the effect is not associated with the catalytic effect of the crushed rock, but with the adsorption of aromatic compounds on the grains of the rock. Aromatic compounds are more prone to adsorption than the saturates. Mainly because of the linear structure of saturates and aromatic rings in aromatics. This effect is possible in the oil reservoir during in-situ combustion too.

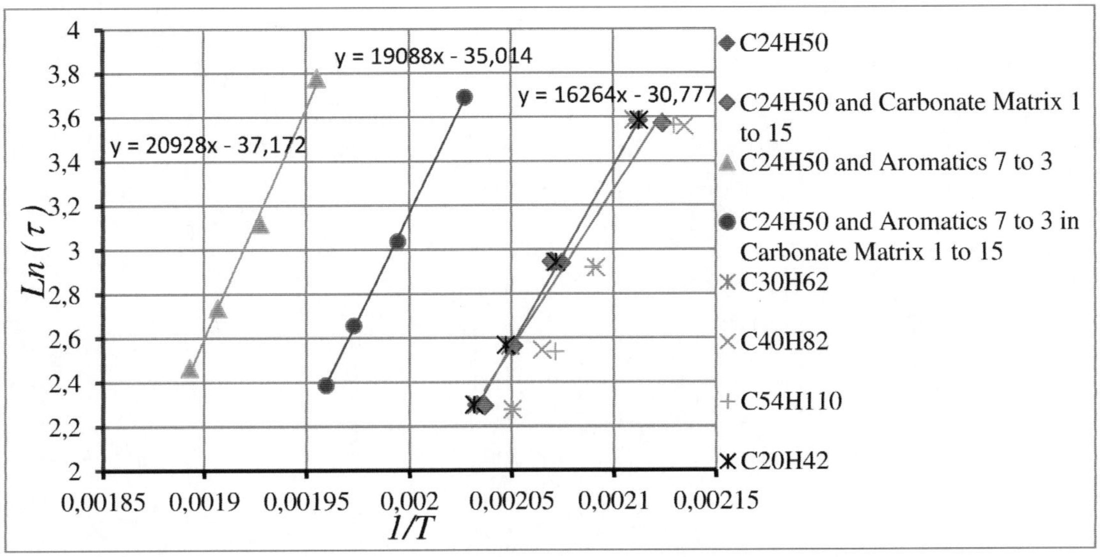

Figure 4 Induction time vs the inverse initial temperature for different saturates, C_{24} with crushed core, C_{24} with aromatic compounds, C_{24} with aromatic compounds and crashed core.

Figure 4 shows the dependence of induction time (period) for all studied compounds and mixtures. Saturates with linear structure, as mentioned earlier, did not significantly differ from each other in oxidation process, the induction period has the same dependence on initial temperature. The crushed rock does not affect the induction period as well (red rhombs), although the presence of rock significantly changes the form of kinetic curves.

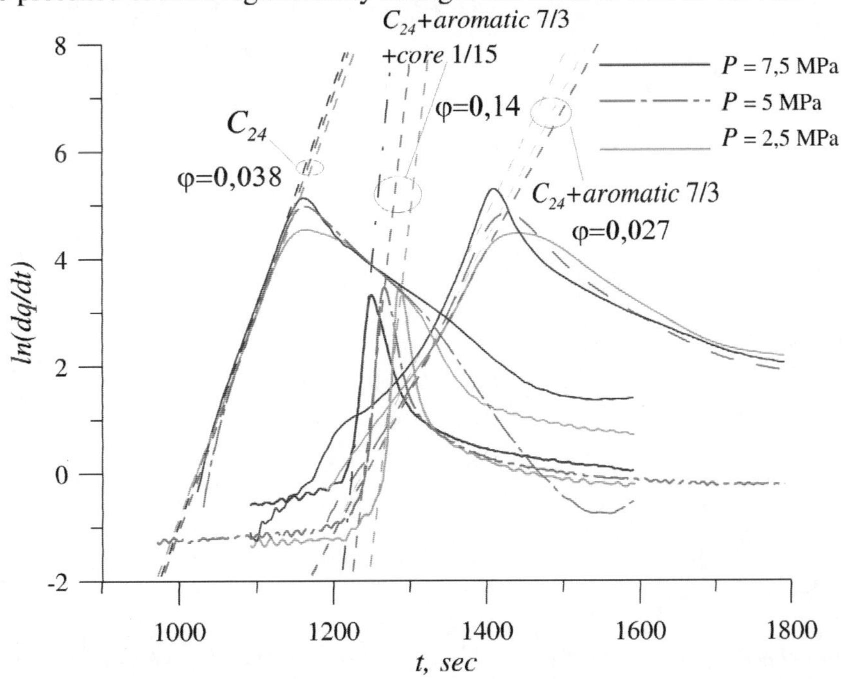

Figure 5 Natural logarithm of kinetic curve opposite time for pure C_{24}, C_{24} with aromatic compounds, C_{24} with aromatic compounds and crashed core under 2.5, 5, 7.5 MPa.

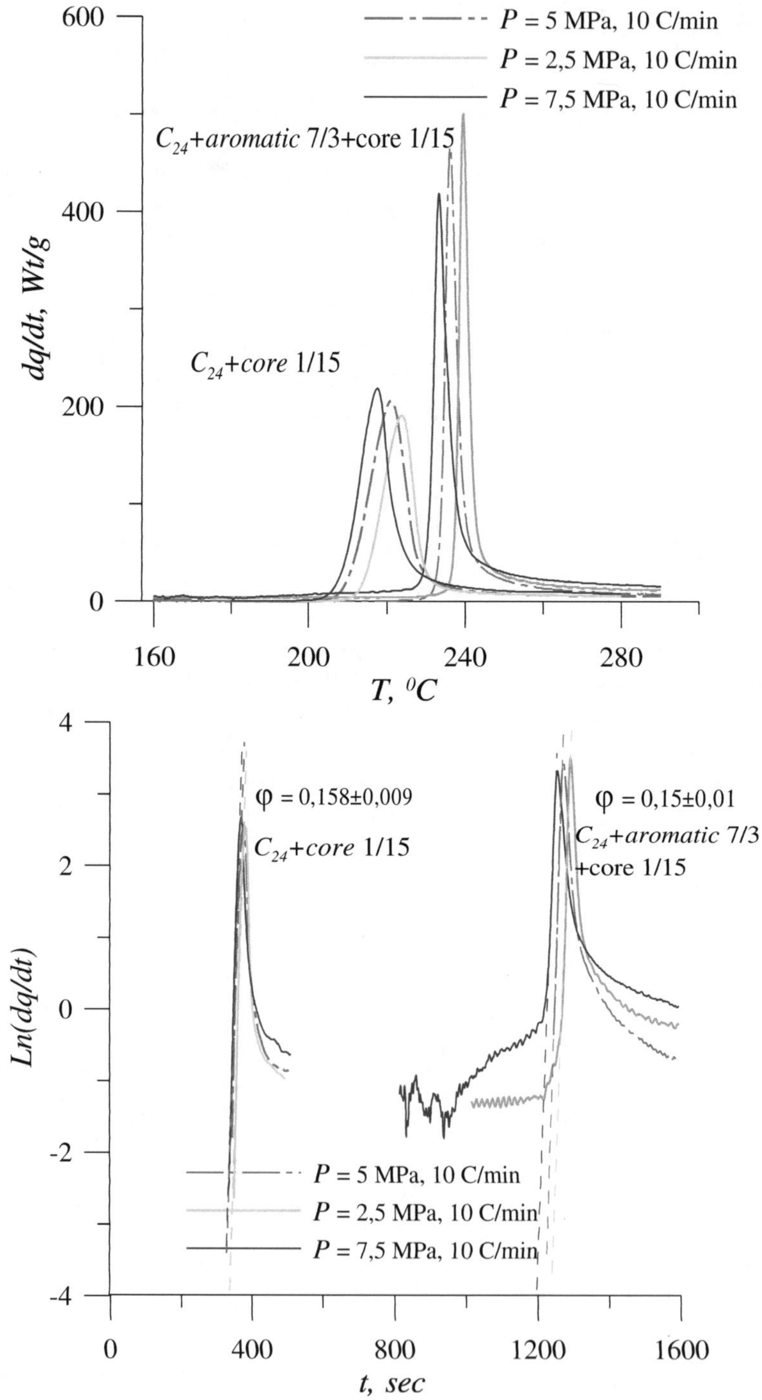

Figure 6 $C_{24}H_{50}$ mixed with aromatics, $C_{24}H_{50}$ mixed with aromatics and crushed core, combustion curves under 2.5, 5, 7.5 MPa (a) and natural logarithm of the same curves opposite time (b).

For tetracosane with aromatics 7/3 (green triangles) one observes the increase in initial temperature and in the slope of the straight line that characterizes the activation energy of the initial stages of oxidation. That is the effect of inhibition. Adding rock material shifts the line to the right (purple circles) and reduces its slope (the activation energy of the process).
Further we examined the influence of crushed rock on the initial stages. The exponential time dependence of initial heat release is expressed as straight lines at the figure 5. The φ factor for mixtures of C_{24} with aromatics on the core, as well as for pure C_{24} on the core within the experimental error reaches the same value regardless of whether there are aromatic compounds in the mixture applied to the crushed rock or not $\varphi_{C24+core}$= 0,158±0,009 and $\varphi_{C24+Arom+core}$= 0,15±0,01. This unexpected result suggests much larger influence of the geometry of the surface, than the inhibition effect of aromatics (figure 5 and 6b). This result also suggests that aromatic compounds are strongly adsorbed on the rock grains and does not take part in the initial stages of oxidation.
This effect requires more detailed study. PDSC experiments let one to determine the limit of adsorption i.e., the minimum concentration of aromatics in the saturates, which support the maximum amount of the adsorbed inhibitors (all the grains covered with adsorbed aromatics). After we reach this concentration of aromatics the effect of acceleration of the initial stages needs to disappear. Additionally, the effect of the surface should depend on the adsorbing properties of rocks, the properties of crushed rock, etc.
Such effect obviously plays a role for application of in-situ combustion in the oilfield. Therefore, to determine the effect of rock the experiments should be performed with the rock, which structure is as close as possible to reservoir internal structure.

6. Pressure effects on the oxidation

Another interesting effect was the influence of pressure on the peak of the oxidation kinetic curves for tetracosane and for tetracosane on the rock (figure 6a). As mentioned above in the case of oxidation in presents of rock the initial heat release and the induction period depend on pressure (table 2). The higher the pressure, the lower the induction period is. Most likely this indicates the presence of free radicals not only in liquid but in gas phase too at the initial stages of oxidation.
For tetracosane the higher the pressure, the more pronounced the peak of heat output is. This is common pressure effect which is in increasing the oxygen concentration.
The presence of aromatics leads to an inverse relationship – the lower the pressure the higher heat peak is (figure 6a). Most likely the low pressure leads to better evaporation of the inactive radicals formed by aromatic compounds. This also causes the decrease of the induction period, the more rapid spending of the inactive radicals. But this effect also requires further researches.

7. Conclusions

The oxidation experiments performed by PDSC with saturates, aromatics and crushed core has shown several effects which should be taken into account when in-situ combustion is considered.

- Saturates are the components which oxidize in low temperature mode. Thus high saturates concentration in crude oil lead to the oxidation mostly in LTO region with small amount of fuel formed. And this is lucky for high pressure air injection application

- The aromatic compounds act as inhibitors of initial oxidation stages. Aromatic additives act in decreasing the initial temperature of oxidation, and result in a longer induction period before combustion. This means that high aromatic concentration in crude oil causes low activity and long period of self ignition for in situ combustion application.
- The pores media always act as the adsorption material for aromatic and oxidized components and accelerate the initial stages of oxidation regardless are there any chemical catalysts in mineral composition or not.
- The pressure dependence of saturates oxidation indicates mainly liquid phase oxidation. In presents of aromatics and crushed core the pressure dependence indicates the gas phase oxidation.

All this results can propagate to the reservoir conditions, but more precise estimation of rock influence it is better to perform experiments with real, not crushed core samples and high pressure.

Acknowledgements

The work has been performed according to the Russian Government Program of Competitive Growth of Kazan Federal University.

References

[1]. Freitag N P 2014 Chemical Reaction Mechanisms That Govern Oxidation Rates During In-Situ Combustion and High-Pressure Air Injection, *SPE 170162*

[2]. Ushakova A, Zatsepin V, Varfolomeev M, Emelyanov D 2017 Study of the Radical Chain Mechanism of Hydrocarbon Oxidation for In Situ Combustion Process *Hindawi Journal of Combustion* ID 2526596, 11 p https://doi.org/10.1155/2017/2526596

[3]. Freitag N P, Exelby D R 2005 A SARA-Based Model for Simulating the Pyrolysis Reactions That Occur in High-Temperature EOR Process *JCPT* Vol 45, N 3

[4]. Sequera B, Moore R G, Mehta S A, Ursenbach M G 2010 Numerical Simulation of In-Situ Combustion Experiments Operated Under Low Temperature Oxidation, *JCPT* Vol 49 N1 p 55-64

[5]. Gutierrez D, Moore R G, Ursenbach M G, Mehta S A 2012 The ABCs of In-Situ-Combustion Simulations: From Laboratory Experiments to Field Scale *SPE- 148754*

[6]. Freitag N P, Verkoczy B 2005 Low-Temperature Oxidation of Oils in Terms of SARA Fractions Why Simple Reaction Models Don't Work *JCPT* Vol 44, N2

[7]. Klinchev V A, Zatsepin V V, Ushakova A S, Telyshev S V 2014 Laboratory studies and implementation of in-situ combustion initiation technology for air injection process in the oil reservoirs *SPE-171244*

[8]. Semenov N N 1943 Advances of Chemical Kinetics in the Soviet Union *Nature* V.151, N 3824 p185-187

IOP Conf. Series: Earth and Environmental Science **155** (2018) 012014 doi:10.1088/1755-1315/155/1/012014

High-resolution NMR study of light and heavy crude oils: "structure-property" analysis

I Rakhmatullin[1], S Efimov[1], M Varfolomeev[2] and V Klochkov[1]

[1] Institute of Physics, Kazan Federal University, 18 Kremlevskaya St., Kazan, 420008 Russian Federation
[2] Department of Physical Chemistry, Kazan Federal University, 18 Kremlevskaya St., Kazan, 420008 Russian Federation

E-mail: ilfat89rakhmatullin@gmail.com

Abstract. Measurements of three light and one heavy crude oil samples were carried out by high-resolution nuclear magnetic resonance (NMR) spectroscopy methods. Quantitative fractions of aromatic molecules and functional groups constituting oil hydrocarbons were determined, and comparative analysis of the oil samples of different viscosity and origin was done.

1. Introduction

^{1}H and ^{13}C NMR spectroscopy is well-known recognized technique for establishing structural formulas, spatial and electronic structure of either first synthesized or isolated from natural raw materials individual organic compounds. In recent years, there has been a growing interest in using NMR spectroscopy for studying various objects in petrochemistry [1]. The NMR method opens wide opportunities in studying the structure of oil disperse systems and determining their physical and chemical characteristics. The NMR method can help to determine aromatic and aliphatic hydrogens and carbons in oil samples. Areas under resonance peaks in NMR spectra characterize the amount of ^{1}H or ^{13}C atoms of corresponding chemical types. Integration of the peaks and subsequent manipulations and calculations yield quantitative proton and carbon type analyses [2]. ^{1}H and ^{13}C NMR can be applied to obtain information on content of general functional groups (tertiary and primary carbon atoms, aromatic cores) and possible presence of olefins or water impurity [3]. However, it has not been yet widely used for analysis of high-molecular weight oil samples because of their complex structural organization. Number of works on the NMR analysis of oils and petroleum products, including oil residues, is limited. This circumstance is largely due to the natural shift of interests towards the study of the properties of heavy oil, for which the possibilities of NMR are limited. In particular, the correlation between viscosity of oil and relaxation times in these objects is no longer observed clearly as for light oils, and the use of pulsed NMR method is becoming increasingly problematic.

High resolution NMR spectroscopy has provided detailed chemical information on the proton and carbon chemistry of petroleum materials for over 60 years [4]. Today high resolution NMR instrumentation used to derive detailed multinuclear hydrocarbon information is based on superconducting magnet technologies.

IOP Conf. Series: Earth and Environmental Science **155** (2018) 012014 doi:10.1088/1755-1315/155/1/012014

In the case of oil and petroleum products containing typically hundreds of compounds, mainly hydrocarbons, an important feature of NMR spectroscopy is a strict correlation of integral intensities of separate groups of signals in certain chemical shifts ranges in ^1H and ^{13}C NMR spectra with the content of the corresponding molecular fragments. Note that widely used in petrochemical methods of analysis approaches such as elemental (X-ray fluorescence, absorption and emission optical spectroscopy), fragment and component (IR spectroscopy, mass spectrometry, chromatography) analyses are indirect (unlike NMR) because they require using reference materials for quantitative investigations.

It should also be noted that the quantitative ^{13}C NMR spectrum is the only direct method for measuring aromaticity (C_{ar}) – relative percentage of aromatic carbon atoms in hydrocarbons.

The aim of this study was to determine the qualitative and quantitative composition of four different crude oil samples by means of the modern methods of high-resolution NMR spectroscopy. ^1H, ^{13}C NMR experiments were applied to provide detailed information on the hydrocarbon chemistry of the chosen petroleum objects.

2. Materials and Methods

Studied samples are presented in Table 1: there are low viscosity oils from Middle-Nazym oilfield (Bazhenov formation) (**1**, **2**) (less than 8 cP), Vishanskoe oilfield (Republic of Belarus) (**3**), and the last one is high viscosity oil from Ashal'cha oilfield (Republic of Tatarstan) (**4**).

Table 1. Studied oil samples.

Sample number	Viscosity, mPa·s	Oilfield
1	7.5	Middle-Nazym (JSC «Ritek», Russia)
2	5.95	Middle-Nazym (JSC «Ritek», Russia)
3	37.2	Vishanskoe (Republic of Belarus)
4	2420	Ashal'cha (Republic of Tatarstan, Russia)

NMR experiments on studied oil samples (**1-4**) were performed on a Bruker Avance II 500 and Bruker Avance III HD 700 (equipped with a CryoProbe) NMR spectrometers. Field lock and shimming were achieved using the deuterium signal from D_2O in a glass capillary placed into the 5 mm NMR tube together with the investigated oil sample. All samples were studied without dilution. ^1H NMR spectra were recorded using 30° pulses (zg30 pulse program); acquisition time was 4.7 s; pre-scan delay was 6.5 µs, and the relaxation delay between scans was 2 s; spectrum width was 12.0 ppm (6000 Hz); 400 scans were accumulated. ^{13}C NMR spectra were recorded using 90° pulses with broadband proton decoupling (zgig pulse program); relaxation delay between pulses was 9 s (and acquisition time was 3.5 s); spectrum width was set to 220.0 ppm; number of scans was 3200. Exponential digital filter with the lb parameter of 10 Hz was applied prior to Fourier transformation. Measurements were made at the temperature of 30°C. All the NMR spectra were integrated after baseline correction, and a mean of minimum three integration values has been taken for each calculation. The relative standard deviation of the results of manual integration did not exceed 3%. Estimation of molar fractions of primary, secondary, tertiary, an aromatic and quaternary carbons was carried out in a way similar to our previous work [3].

3. Results and Discussion

As a comparison of the higher resolution of ^{1}H and ^{13}C NMR experiments performed on high-resolution NMR systems, spectra of the sample 1 obtained on two spectrometers are presented in Figures 1 and 2. More peaks are resolved at 700 MHz but the general spectral view can be seen to be the same. The content of several well-resolved signals indicates the presence of branched aliphatic structures. Typical assignment intervals of ^{1}H and ^{13}C NMR spectra of samples are given in our previous work [3].

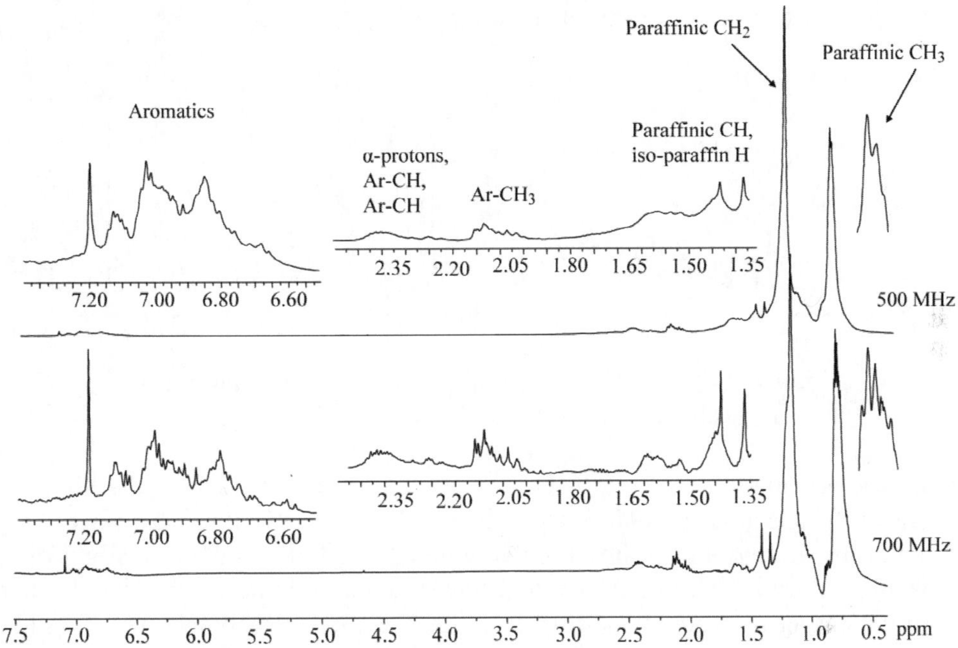

Figure 1. ^{1}H NMR spectra of sample **1** at resonance frequencies 500 MHz and 700 MHz. Observed proton types are indicated on the spectrum.

IOP Conf. Series: Earth and Environmental Science **155** (2018) 012014 doi:10.1088/1755-1315/155/1/012014

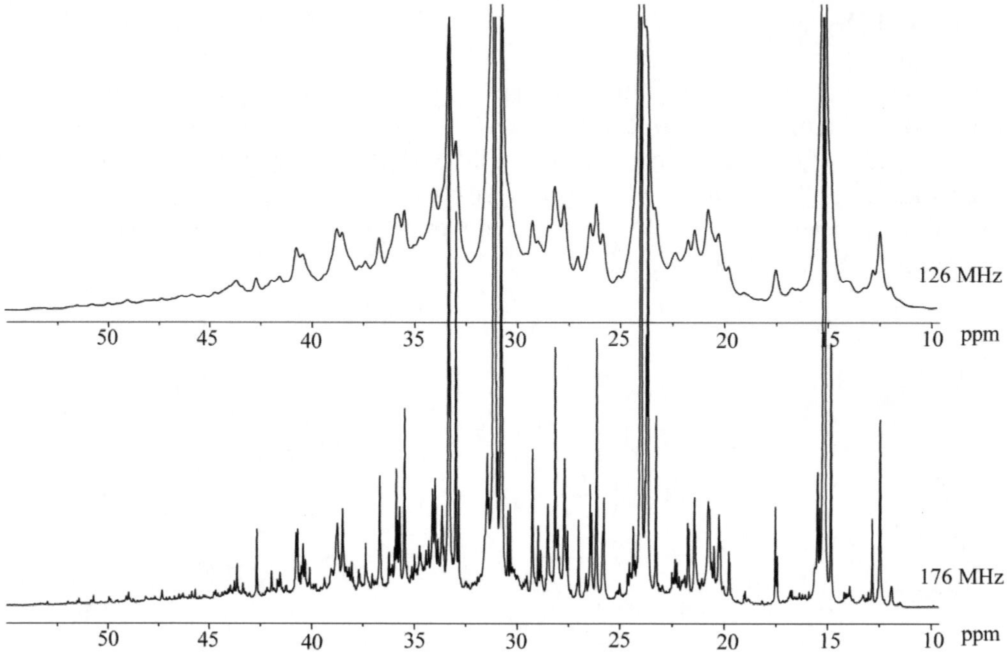

Figure 2. ^{13}C NMR spectra (aliphatic region) of sample **1** at resonance frequencies 126 MHz and 176 MHz.

There are characteristic ranges of aliphatic and aromatic proton signals in the high- and low-fields regions of the spectra. For samples **1** and **2**, similar spectra with well-resolved signals in aliphatic and aromatic regions are observed (Figures 3, 4). For sample **3**, a slight broadening of the signals is observed, while for sample **4** the broadening is more prominent. Such a line broadening, indicating a strong viscosity, leads to the fact that some low-intensity signals merge with the baseline. For example, the signal at 17.5 ppm in ^{13}C NMR spectrum (Figure 4) is clearly visible for samples **1** and **2**, is weakly noticeable for sample **3**, and is not visible for sample **4**.

^{13}C NMR spectroscopy proved to be much more efficient than ^{1}H NMR in the study of oil samples, since the broadening of the signals here has a lesser effect on the information content of the spectra due to a much wider chemical shifts range of ^{13}C [1]. In addition, ^{13}C NMR spectra allow determining the content of quaternary carbon atoms and functional groups not containing hydrogen atoms.

Information obtained by quantitative integration of signals in individual spectral ranges is represented by the fraction of the corresponding carbon atoms relative to their total number:

$$C_i = \frac{I_i}{\sum_j I_j} .$$

It is impossible to obtain clear information on the content of hydrocarbons (alkanes, cyclanes) from ^{13}C NMR spectra, although this information is contained in the fragmentary composition, which can be determined with high accuracy [2].

Table 2 shows the results of estimating the molar content of various carbon groups made by integration of the corresponding regions of ^{13}C NMR spectra.

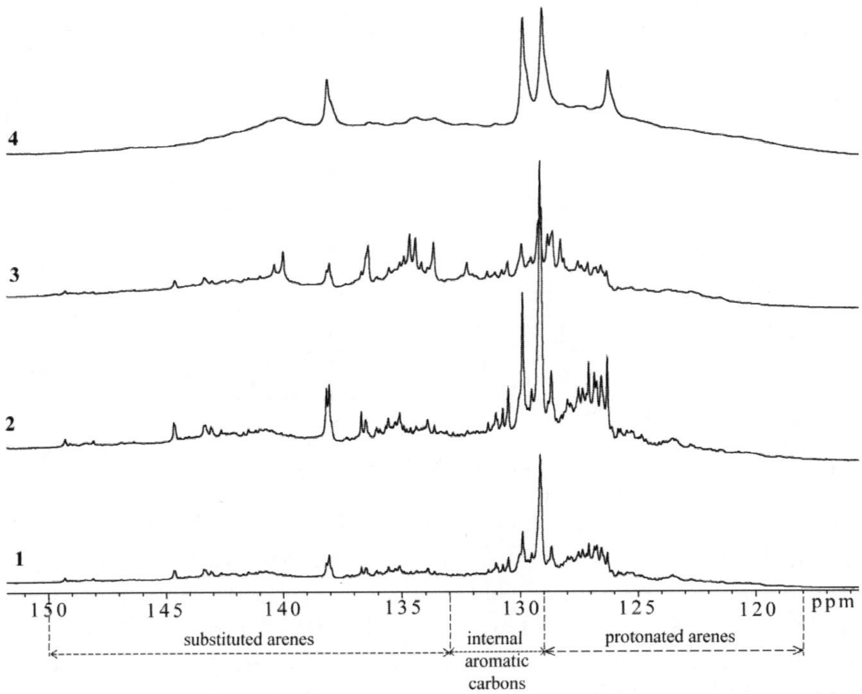

Figure 3. The aromatic region of ^{13}C NMR spectra (176 MHz) of samples **1-4**.

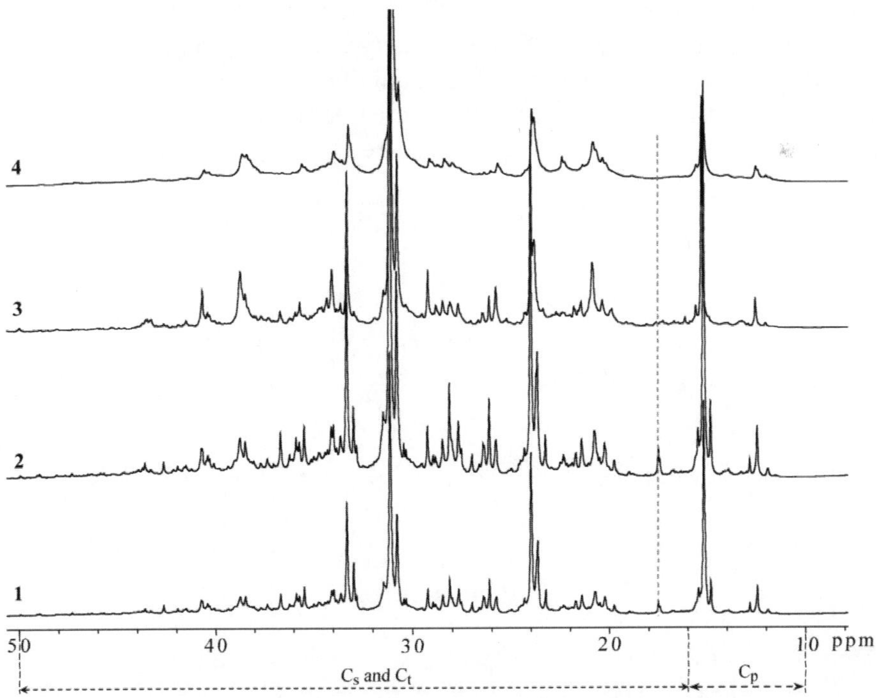

Figure 4. The aliphatic region of ^{13}C NMR spectra (176 MHz) of samples **1-4**. C_p, C_s, C_t stand for primary, secondary, and tertiary carbons.

IOP Conf. Series: Earth and Environmental Science **155** (2018) 012014 doi:10.1088/1755-1315/155/1/012014

Table 2. Molar fractions (%) of primary (C_p), secondary + quaternary (C_{sq}), tertiary (C_t), aromatic (C_{ar}) groups and mean chain length of aliphatic hydrocarbons based on ^{13}C NMR spectra of samples **1-4**.

Group type	Sample number			
	1	2	3	4
C_p	34.7	33.4	25.8	21.6
C_{sq}	48.4	53.8	53.0	48.9
C_t	12.1	4.8	10.0	12.8
C_{ar}	6.4	6.9	12.5	19.3
C_{arom} / C_{aliph}	0.067	0.075	0.141	0.232
mean chain length	4.9	5.1	6.0	6.5

Based on obtained data (Table 2), we can see that the highest percentage of primary and secondary carbons is in sample **2**, which is however almost twice as low in the content of tertiary carbons. Also, twofold and threefold excess of the relative content of aromatic carbon atoms in samples **3** and **4**, respectively, compared with samples of light oils **1** and **2** is observed. So, in the total comparison of aromatic/aliphatic content, the smallest value belongs to sample **1** – 0.067. Mean chain length values turned out to be rather low, which indicates the approximate nature of our model, but there is an obvious tendency to increase the chain length as the viscosity of sample increases.

Evidently, molar fraction of aromatic groups (C_{ar}) (aromaticity) changed the most while the molar fraction of aliphatic groups (C_p, C_{sq}, C_t) changed not so prominently in passing from light oils (**1-3**) to heavy one (**4**). Also there is a mutually inverse dependency for primary and aromatic carbons as the viscosity of the samples increases (Figure 5).

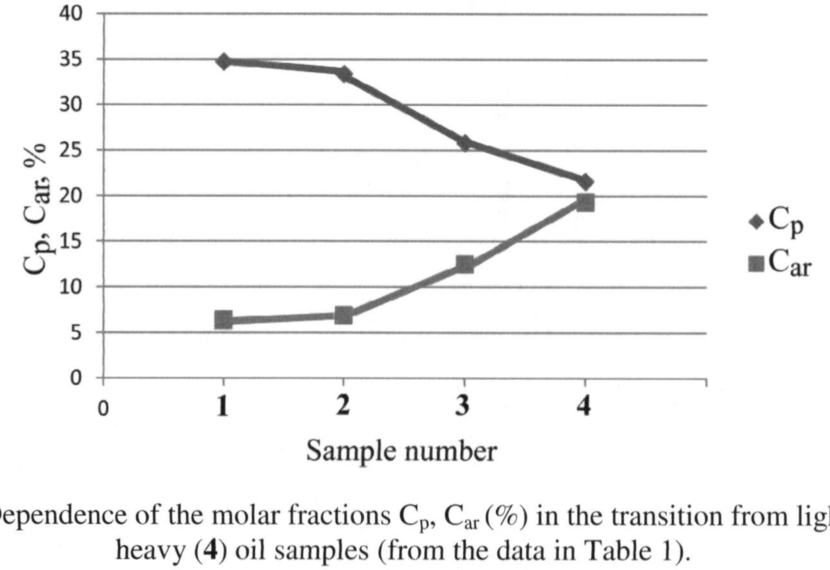

Figure 5. Dependence of the molar fractions C_p, C_{ar} (%) in the transition from light (**1-3**) to heavy (**4**) oil samples (from the data in Table 1).

IOP Conf. Series: Earth and Environmental Science **155** (2018) 012014 doi:10.1088/1755-1315/155/1/012014

4. Conclusions

The samples of some light paraffinic and heavy aromatic oils were studied by high resolution NMR spectroscopy. Aromatic and aliphatic oil components can be distinguished by ^{1}H, ^{13}C NMR analysis. High resolution NMR instrumentation with field strengths of 11.7 Tesla (^{1}H resonances frequencies of 500 MHz) is adequate for most liquid-state petroleum applications as the higher resolving power at very high fields does not provide any significant extra discernment of detailed chemistry. Using NMR, one cannot absolutely settle the problem of the identification and evaluation of hydrocarbon composition in the crude fluid, and it should be considered in combination with existing traditional methods for studying the oil composition. However, this technique can enrich oil researches and improve its application efficiency.

References

[1] Kalabin G, Kanitskaya L and Kushnarev D 2000 *Quantitative NMR spectroscopy of natural organic raw materials and its processing products* (Moscow: Chemistry) [in Russian]
[2] Kushnarev D F *et al* 1989 *Petrol. Chem. U.S.R.R.* **29** pp 149–159
[3] Rakhmatullin I, Efimov S, Margulis B and Klochkov V 2017 *Journal of Petroleum Science and Engineering* **156** pp 12–18
[4] Edwards J 2011 *A review of applications of NMR spectroscopy in the petroleum industry. Chapter 16 in "Spectroscopic analysis of petroleum products and lubricants"* (USA: ASTM International) pp 423-473

The work has been performed according to the Russian Government Program of Competitive Growth of Kazan Federal University

Numerical solving of highly viscous fluids filtration in porous media for nonlinear filtration laws with power growth

I B Badriev[1], N V Kalacheva[2], A I Shangaraeva[3] and V A Sudakov[3]

[1] Institute of Computational Mathematics and Information Technologies, Kazan Federal University, Kremlevskaya str., 18, Kazan 420008 Russian Federation
[2] Lobachevsky Institute of Mathematics and Mechanics, Kazan Federal University, Kremlevskaya str., 18, Kazan 420008 Russian Federation
[3] Institute of Geology and Petroleum Technologies, Kazan Federal University, Kremlevskaya str., 18, Kazan 420008 Russian Federation
E-mail: ildar.badriev1@mail.ru

Abstract. We study the steady-state filtration process of an incompressible high-viscosity fluid follows the nonlinear filtration law. The generalized statement of this problem is formulated in the form of an operator equation with a monotone operator in a Banach space. To solve this operator equation, we propose an iteration method that does not require the inversion of the original operator. Each step of the iterative process reduces to solving the boundary value problem for the Laplace equation. In the Matlab environment, a software complex was developed, with the help of which numerical calculations were performed for model filtration problems. The analysis of numerical results is carried out.

1. Introduction

The specialists are interested in rational development of oil fields, aimed at minimizing the number of wells, as their increase leads to increased environmental pollution. The specificity of a viscoplastic fluid consists in the fact that it remains immobile in a porous medium if the pressure gradient modulus does not exceed a certain limiting value (limiting gradient). During the filtration, areas of immobile oil (stagnant zones) may arise. Determining the boundaries of such zones is undoubtedly an urgent problem.

In this paper, we study the process of steady-state filtration of an incompressible high-viscosity fluid follows a nonlinear filtration law with a limiting gradient (see, e.g., [1–7]). We assume that the function corresponding to this law is continuous, non-decreasing and has power growth, not necessarily linear, on sets corresponding to the flow zone. The generalized statement of this problem is formulated in the form of an operator equation in a Banach space. Under the conditions formulated above on the function that defines the filtration law, the operator in this equation is monotone, continuous, and coercive [8]. These properties of the operator allow us to use the general results of the theory of monotone operators [8–10] to prove the solvability of the equation. To solve the operator equation, an iterative method is proposed that does not require the inversion of the original operator. Each step of the iterative process reduces to solving the boundary value problem for the Laplace equation. Under additional conditions on the function describing the filtration law, the convergence of the process is established. The proposed method was implemented numerically. The results of numerical experiments are carried out for model problems. These results confirmed the effectiveness of the proposed iteration method. The behavior of the boundaries of stagnant zones is studied depending on the exponent in the filtration law function.

We note that exact solutions are known for a number of filtration area and filtration laws (when the filtration law function has linear growth in the flow zone) (see, e.g., [1–3]). However, in the case of arbitrary filtration area (including three-dimensional ones) and an arbitrary degree of growth in the filtration law function, the solving of the seepage problems is possible only using approximate methods.

2. The problem statement

We consider the problem of determining the stationary pressure fields p and the filtration rate $w=(w_1,w_2)$ of a high-viscosity fluid in the domain Ω of the plane R^2 satisfying the continuity equation and the effective nonlinear filtration law with a limiting gradient and the corresponding boundary conditions

$$\text{div } w(x)=f_*(x), w(x)=\frac{g(|\nabla p(x)|)}{|\nabla p(x)|}\nabla p(x), x\in\Omega,\ p(x)=0, x\in\Gamma_1, (w,\mathbf{n})=0, x\in\Gamma_2\ , \quad (1)$$

where $\Gamma=\Gamma_1\cup\Gamma_2$, Γ is the boundary of the domain Ω, $\mathbf{n}$ is the unit outer normal to Γ vector, f_* is the density of external sources, and g is the function that determines a filtration law. We assume that the function g satisfies the following conditions (see Figure 1):

$$g \text{ is continuous, does not decrease, } g(\xi)=0 \text{ for } \xi\le\beta, \quad (2)$$

here $\beta\ge 0$ is a limiting gradient; there exist $c_0>0$, $c_1>0$, $c_2>0$, $p\ge 2$ such that

$$c_0\xi^{p-1}-c_1\le g(\xi)\le c_2\xi^{p-1} \text{ for } \xi\ge\beta. \quad (3)$$

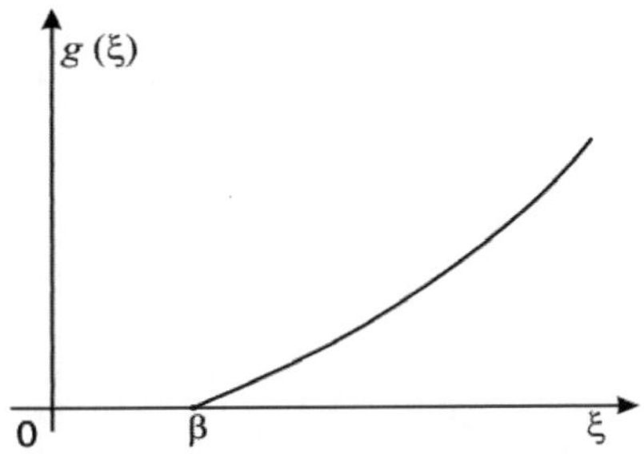

Figure 1. Nonlinear power filtration law

We define the operator $G:R^2\to R^2$ by the formula $G(y)=g(|y|)y/|y|$, $y\ne 0$, $G(0)=0$. Let's $V=\{\eta\in W_p^{(1)}(\Omega):\eta(x)=0, x\in\Gamma_1\}$ is the Sobolev space, f_* generate a linear continuous functional on V. By virtue of conditions (2), (3), the form $a(u,\eta)=\int_\Omega (G(\nabla u)(\nabla u,\nabla\eta)dx$ generates the operator $A:V\to V^*=W_{p^*}^{(-1)}(\Omega)$, $p^*=p/(p-1)$, by the formula $\langle Au,\eta\rangle=a(u,\eta)$ (see [11]), where $\langle\cdot,\cdot\rangle$ is the duality relation between V and V^*, V^* is the space conjugate to V (see [9]). By the solution of the filtration problem (1) we mean a function $u\in V$ satisfying equation

$$Au=f\ , \quad (5)$$

where the element $f\in V^*$ is determined by the formula $\langle f,\eta\rangle=\int_\Omega f_*\eta\,dx$. The operator A is continuous, monotone, and coercive; therefore, equation (1) has at least one solution (see, e.g., [8–10]). We note that in [12–16] the filtration problems with the multivalued filtration law in the case $p=2$ are studied. The filtration problem was formulated in the form of a variational inequality.

3. Iterative method and numerical experiments.

To solve the operator equation (5), we use a two-layer iterative process of the form [17–20]

$$-\Delta(u^{k+1}-u^k)=\tau_k(Au^k-f),\quad k=0,1,2,\ldots, \quad (6)$$

where $\tau_k>0$ are the iteration parameters, u^0 is given element. If the function g, in addition to (2), (3) satisfies also condition (see. [11, 17]) $(g(\xi)-g(\zeta))/(\xi-\zeta)\le c_3(1+\xi+\zeta)^{p-2}$, $c_3>0$ and $\tau_k=\min\{1,1/(\alpha+\mu_k)\}$, $\mu_k=\mu(\|u^k\|+\|Au^k-f\|_{V^*}^{-1})$, , where $\mu(\xi)=c_4(1+2\xi)^{p-2}$; $c_4>0$; α is an arbitrary positive number, then the iterative sequence $\{u^k\}$ constructed according to (6) is bounded; any weakly limit point of the sequence $\{u^k\}$ is a solution of problem (5) [18]. A study of the convergence of the method is based on the proof of the quasi-potentiality and bounded Lipschitz continuity [9, 15] of the operator A: for all $u, \eta \in V$, the inequality $\|Au - A\eta\|_{V^*} \le \rho(R)\Phi(\|u-\eta\|)$ holds, where $R = \max\{\|u\|, \|\eta\|\}$, ρ is the non-decreasing on $[0,+\infty)$ function; Φ is the increasing on $[0,+\infty)$ function, $\Phi(0)=0$, $\Phi(\xi) \to +\infty$ for $\xi \to +\infty$. In this case, in the two-layer iterative method considered in [6, 16], as a pre-conditioner, generally speaking, is the nonlinear duality operator [8], generated by the function Φ. For the filtration problem studied in the present paper, the operator A generated by the form (4) satisfies the condition of bounded Lipschitz continuity with functions $\rho = \mu$ and $\Phi(\xi) = \xi$, and the Laplace operator acts as a pre-conditioner, and the iterative process is written in the form (6).

A software package in the MatLab environment is developed. Numerical experiments for model filtration problems are carried out. The dependence of the boundaries of stagnant zones, i.e., the sets in the filtration area, on which the modulus of the pressure gradient is less than the limiting one and the flow is absent, on the degree of p in the filtration law was investigated.

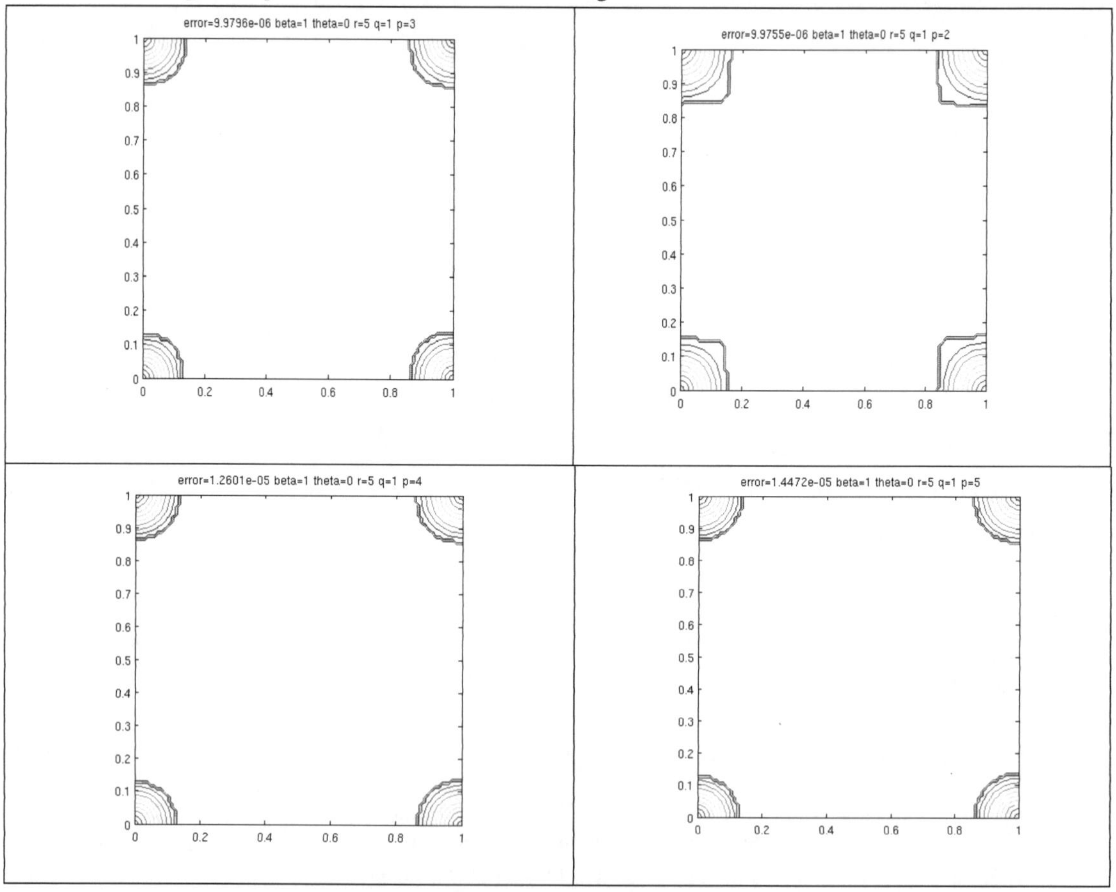

Figure 2. The boundaries of the stagnant zones at a values of p: a) 2; b) 3; c) 4; d) 5.

The results of numerical experiments, i.e., the boundaries of stagnant zones, are shown in Fig. 2. The filtration area was a unit square, in the center of which there is a well with a production rate q = 1, we put $\beta = 1$, $g(\xi) = (\xi - \beta)^{p-1}$ for $\xi \ge \beta$. It should be noted that for $p = 2$ (the case of a Hilbert space), the results (see Fig. 2a) correspond to the results obtained by the splitting method [21–25]. As can be seen from Fig. 2, with increasing p, the stagnant zones decrease, which is explained by the increase in the flow intensity in the flow zones. The results obtained correspond to the expected flow pattern.

Acknowledgement

The work is performed according to the Russian Government Program of Competitive Growth of Kazan Federal University and with financial support of the RFBR and the Government of the Republic of Tatarstan within the framework of the scientific project No. 15-41-02315.

References

[1] Dmitriev N M., Maksimov V M and Mamedov M T 2010 Laws of flow with a limiting gradient in anisotropic porous media *Fluid Dynamics* **45** (2) 223–9.

[2] Bernardiner M G and Entov V M 1975 *Hydrodynamic Theory of Anomalous Fluid Flow through Porous Media* (Nauka: Moscow) (in Russian).

[3] Mirzajanzade A Kh and Entov V M 1985 *Hydrodynamics in Drilling* (Nedra, Moscow) (in Russian).

[4] Dmitriev N M, Maksimov V M and Ryabchukov E A 2006 Laws of Viscoplastic Fluid Flow in Anisotropic Porous and Fractured Media *Fluid Dynamics* **41**(4) 585–92.

[5] Basniev K S, Dmitriev N M, Kanevskaya R D and Maksimov V M 2006 *Subterranean Hydromechanics* (Institute of Computer Investigations, Moscow & Izhevsk) (in Russian)

[6] van Golf-Racht T D 1982 *Fundamentals of Fractured Reservoir Engineering (Developments in Petroleum Science)* (Amsterdam: Elsevier).

[7] Badriev I B and Nechaeva L A 2013 Mathematical simulation of steady filtration with multivalued law *PNRPU Mechanics Bulletin* **3** 37–65.

[8] Lions J.-L 1969 *Quelque problèmes méthodes de résolution des problèmes aux limites nonlinéaires* (Dunod: Paris).

[9] Gajewskii H, Gröger K and Zacharias K. 1974 *Nichtlineare Operatorgleichungen und Operatordifferentialgleichungen* (Akademie-Verlag: Berlin).

[10] Ekeland I and Temam R 1976 Convex Analysis and Variational Problems (Amsterdam: North-Holland).

[11] Lyashko A D and Karchevskii M M 1983 Difference Methods of Solving Nonlinear Problems in the Theory of Filtration *Soviet Mathematics* **27** (7) 34–56.

[12] Badriev I B 2013 On the solving of variational inequalities of stationary problems of two-phase flow in porous media *Applied Mechanics and Materials* 392 183–7.

[13] Badriev I B and Nechaeva L A 2013 Mathematical simulation of steady filtration with multivalued law *PNRPU Mechanics Bulletin* **3** 37–65

[14] Badriev I B and Karchevskii M M (1989) Convergence of the iterative Uzawa method for the solution of the stationary problem of seepage theory with a limit gradient *Journal of Soviet Mathematics* **45** (4) 1302–09

[15] Badriev I B, Zadvornov O A and Saddek A M 2001 Convergence Analysis of Iterative Methods for Some Variational Inequalities with Pseudomonotone Operators *Differential Equations* **37** (7) 934–42

[16] Badriev I B and Fanyuk B Y 2012 Iterative methods for solving seepage problems in multilayer beds in the presence of a point source *Lobachevskii Journal of Mathematics* **33** (4) 386–99

[17] Badriev I B and Karchevskii M M 1994 Convergence of an iterative process in a Banach space *Journal of Mathematical Sciences* **71** (6) 2727–35

[18] Badriev I B, Zadvornov O A and Lyashko A D 2004 A Study of Variable Step Iterative Methods for Variational Inequalities of the Second Kind *Differential Equations* **40** (7) 971–83

[19] Badriev I B, Makarov M V and Paimushin V N 2016 Numerical Investigation of Physically Nonlinear Problem of Sandwich Plate Bending *Procedia Engineering* **150** 1050–55

[20] Badriev I B, Garipova G Z, Makarov M V and Paymushin V N 2015 Numerical solution of the issue about geometrically nonlinear behavior of sandwich plate with transversal soft filler *Research Journal of Applied Sciences* **10** (8) 428-35

[21] Badriyev I B, Zadvornov O A, Ismagilov L N and Skvortsov E V 2009 Solution of plane seepage problems for a multivalued seepage law when there is a point source *Journal of Applied Mathematics and Mechanics* **73** (4) 434–42

[22] Badriev I B 1989 Application of duality methods to the analysis of stationary seepage problems with a discontinuous seepage law Journal of Soviet Mathematics 45 (4) 1310–14

[23] Badriev I B and Zadvornov O A 2003 A decomposition method for variational inequalities of the second kind with strongly inverse-monotone operators *Differential Equations* **39** (7) 936–944.

[24] Badriev I B 1989 Regularization of the nonlinear problem of seepage theory with a discontinuous law *Journal of Soviet Mathematics* **44** (5) 681–91

[25] Badriev I B, Banderov V V, Pankratova O V and Shangaraeva A I 2016 Mathematical simulation of a steady process of anisotropic filtration *IOP Conference Series: Materials Science and Engineering*, **158** (1), art. 012013

IOP Conf. Series: Earth and Environmental Science **155** (2018) 012016 doi:10.1088/1755-1315/155/1/012016

Analytical Model of Steam Chamber Evolution from Vertical Well

D V Shevchenko[1,2], S A Usmanov[2], A I Shangaraeva[2] and T A Murtaizin[2]

[1] Kazan Innovative University named after V.G.Timiryasov, Kazan, Russia
[2] Institute of Geology and Petroleum Technologies, Kazan Federal University, Kazan, Russia

E-mail: sausmanov@gmail.com

Abstract. This paper is aimed to check the possibility of applying the Steam Assisted Gravity Drainage in vertical wells. This challenge seems to be vital because most of the natural bitumen reservoirs are found to occur above the oil fields being developed so that a well system is already available at the stage of field management. The existing vertical wells are hard to be used for horizontal sidetracking in most of cases as the bitumen reservoir occurs at a shallow depth. The matter is to use the existing wells as vertical ones. At the same time, it is possible to drill an additional sidetrack as a producer or an injector.

1. Introduction

Steam assisted gravity drainage (SAGD) [1, 2] is one of the most successful thermal methods for heavy oil recovery [1, 2]. Well-known approximate SAGD methods are described by R.M. Butler [3, 4]. The vast majority of papers describe the development of a heavy oil field by a system of horizontal wells since this approach is the most technologically advanced [5, 6]. For such systems, simple integral estimates of the steam chamber parameters have been determined [7, 8].

This paper uses the methods based on integral correlations [9] to study how the steam chamber develops from a vertical well. The approach is based on the total balance correlations that result in an exact relationship for the integral characteristics, a weak dependence of the qualitative results on the local disturbances of the flow parameters and the shape of the steam chamber [8]. The approach makes it possible to get qualitative conclusions without performing sophisticated numerical calculations. For a pair of horizontal wells, it is shown in [8] how the results of a similar approach match well with the results of numerical simulation performed on a full-size hydrodynamic simulator.

2. Methodology

Test schematic is shown in Fig. 1. The assumption is made that the steam chamber is likely a circular cone with a constant height h and a variable radius R. The height h is determined primarily by the reservoir thickness, as well as by the distance to the production well h_{min}, which is necessary to prevent steam breakthrough.

IOP Conf. Series: Earth and Environmental Science **155** (2018) 012016 doi:10.1088/1755-1315/155/1/012016

The production rate is assumed to be constant and even q. Such an assumption is entirely permissible within the integral model for the rates that do not vary greatly with time.

In all formulas, the index r stands for the reservoir, the index t stands for the top boundary, the index o stands for the oil, w stands for the water, s stands for the steam. The index T will stand for the heat flux.

It appeals to qualitatively estimate the volume V and typical dimensions R, l of the steam chamber development from the vertical well, the measure of the produced oil q_o and the pattern of how the oil ratio changes in the production q_o/q, where q is the total production rate of the production well.

Let q be the volume flow rates of fluids per unit time.

The total flow rate is equal to the flow of water q_w and oil (bitumen) q_o:

$$q = q_w + q_o. \tag{1}$$

The volume of the produced water is:

$$q_w = q_s + q_{wr} - q_{sr} \tag{2}$$

where q_s is the volume of water injected as steam, q_{wr} is the volume of reservoir water swept, q_{sr} is the volume of water in the form of steam, which fills the steam chamber (all volumes are referred to a unit of time).

The formulas (1), (2) yield:

$$q_s = q - q_o - q_{wr} + q_{sr} \tag{3}$$

The last two terms are valid:

$$q_{wr} = m \cdot \Delta s_w \frac{dV}{dt} \tag{4}$$

$$q_{sr} = \frac{\rho_s}{\rho_w} m \cdot \Delta s_m \frac{dV}{dt} \tag{5}$$

where ρ is the density, m is the porosity, V is the volume of the steam chamber, t is the time, $\Delta s_w = s_{w0} - s_{w\min}$, $\Delta s_m = 1 - s_{w\min} - s_{o\min}$, s_{w0}, $s_{w\min}$, s_{o0}, $s_{o\min}$ are initial and residual water saturation and oil saturation.

The change in the volume of the steam chamber is proportional to the oil production rate:

$$q_o = m\Delta s_o \frac{dV}{dt}, \tag{6}$$

where $\Delta s_o = s_{o0} - s_{o\min}$.

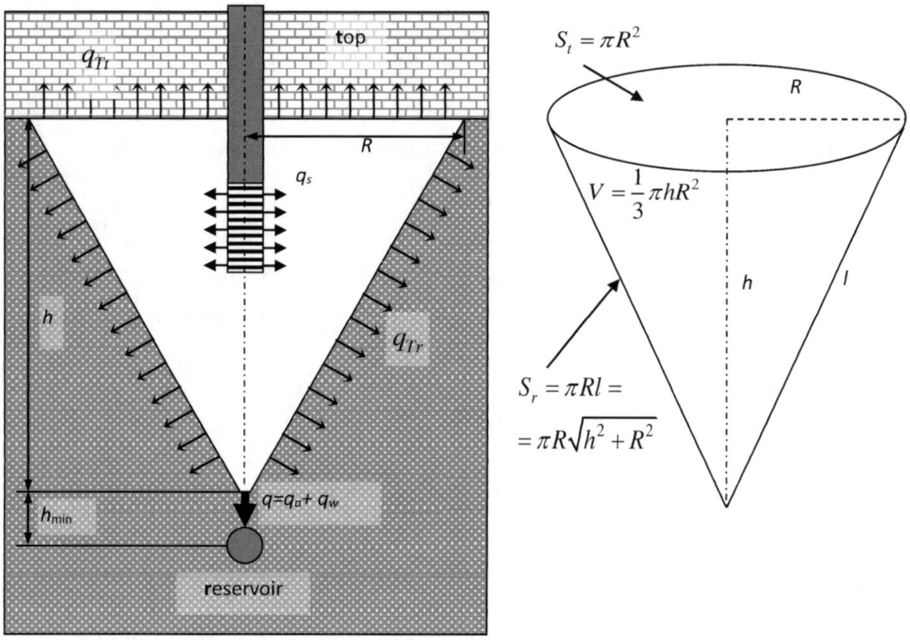

Fig. 1 Steam chamber in bitumen reservoir. Fluid and heat fluxes and the correlations between geometric characteristics

Then the total oil production is

$$Q_o = \int_0^t q_o dt = m\Delta s_o \left(V - V_0\right) = m\Delta s_o \left(\frac{1}{3}\pi h R^2 - V_0\right), \tag{7}$$

where V_0 is the volume of the steam chamber formed during preheating t_0, before the bitumen production starts.

The formulas (3)-(7) result in:

$$\frac{q_s}{q} = 1 - \frac{q_o}{q}\left(1 + \frac{\Delta s_w}{\Delta s_o} - \frac{\rho_s}{\rho_w}\frac{\Delta s_m}{\Delta s_o}\right) \tag{8}$$

Let's write the energy mass balance equation of the energy entering the reservoir. The energy that releases during steam condensation heats up the steam chamber and the heat loss to the reservoir and to the top boundary:

$$L\left(q_s \rho_w \varphi - m\rho_s \Delta s_m \frac{dV}{dt}\right) = k_r C_{vr} dT \frac{dV}{dt} + S_t q_{Tt} + S_r q_{Tr}, \tag{9}$$

where L is the latent heat of vaporization of water, φ is the mass fraction of dry steam in the vapor, $dT = T_s - T_0$ is the difference between the temperature of the condensing vapor and the initial reservoir temperature, C_{vr} is the volumetric heat capacity of rocks, S_t, S_r are the areas of steam chamber contact with the top boundary and the reservoir, respectively, q_{Tt}, q_{Tr} are the average heat flux through unit of the corresponding area.

The average flow through the surface of contact with the top boundary is well defined by the formula [10]:

$$q_{Tt} \approx \lambda_t \frac{dT}{b_t\sqrt{a_t(t-t_0)}} \tag{10}$$

where λ_t is the thermal conductivity of the top boundary, a_t is the thermal diffusivity of the top boundary ($a_t = \lambda_t / C_{vt}$), b_t is the coefficient within $0.7 \div 1.5$.

Determination of the average flux through the contact surface with the reservoir required solution to a supplementary homogeneous problem of simulation resulting in an approximate dependence:

$$q_{Tr} \approx \lambda_r \frac{dT}{b_r\left(a_r(t-t_0)\right)^{0,312}} \tag{11}$$

where λ_r is the thermal conductivity of the reservoir, a_r is the thermal diffusivity of the reservoir, b_r is the coefficient within $0.05 \div 0.1$.

The (6)-(11) and geometric relations for areas yield:

$$L\left(\rho_w\varphi\left(1-\frac{q_o}{q}\left(1+\frac{\Delta s_w}{\Delta s_o}-\frac{\rho_s}{\rho_w}\frac{\Delta s_m}{\Delta s_o}\right)\right)-\rho_s\frac{\Delta s_m}{\Delta s_o}\frac{q_o}{q}\right)=C_{vr}dT\frac{1}{m\Delta s_o}\frac{q_o}{q}+ \\ +\frac{1}{q}\pi R^2\lambda_t\frac{dT}{b_t\sqrt{a_t(t-t_0)}}+\pi R\sqrt{h^2+R^2}\lambda_r\frac{dT}{b_r\left(a_r(t-t_0)\right)^{0,312}} \tag{12}$$

The formula of the cone volume produces:

$$R=\sqrt{3V/\pi h}=\sqrt{3\left(V_0+\left(Q_0/m\Delta s_o\right)\right)/\pi h} \tag{13}$$

The formula (12) with the substitution (13) is a differential equation to determine the function $Q_o(t)$. The equation contains the first derivative of the desired function $q_o(t)=Q_o'(t)$, which can be expressed explicitly. The initial condition for this equation is written as:

$$Q_o(0)=0. \tag{14}$$

Let's reduce the system (12)-(14) to the dimensionless form:

$$Ay'=1-B_t\frac{(y_0+y)}{\sqrt{\tau-\tau_0}}-B_r\frac{\sqrt{(y_0+y)(H+y_0+y)}}{(\tau-\tau_0)^{0.312}}, \tag{15}$$

$$y(0)=0, \tag{16}$$

where $\tau = t/t_x$ is the dimensionless time; $y(\tau)=Q_o/qt_x$ is the dimensionless value of accumulated oil recovery; $y'(\tau)=q_o/q$ is the share of oil in the production

$$A = 1 + \frac{\Delta s_w}{\Delta s_o} + \frac{1-\varphi}{\varphi}\frac{\Delta s_m}{\Delta s_o}\frac{\rho_s}{\rho_w} + \frac{C_{vr} dT}{L\rho_w \varphi m \Delta s_o}, \quad B_t = \frac{3b_t dT\sqrt{\lambda_t C_{vt} t_x}}{L\rho_w \varphi h m \Delta s_o}, \quad B_r = \frac{3b_r dT\sqrt{\lambda_r C_{vr} t_x}}{L\rho_w \varphi h m \Delta s_o},$$

$$H = \frac{\pi h^3}{3qt_x} m\Delta s_o, \quad y_0 = \frac{V_0 m \Delta s_o}{qt_x},$$

The typical time t_x is chosen from the proportionality of the dimensionless constants, A, B_t and B_r. Lesser times produce the initial moment of growth of the chamber. Larger times yield a quasi-steady growth typical of an asymptotic equation:

$$Ay' = 1 - B_r \frac{y}{\tau^{0.312}},$$

the solution of which can be determined in quadratures [11]:

$$y = \frac{1}{A}\int_0^\tau \exp\left(\frac{B}{0.688A}\left(\xi^{0.688} - \tau^{0.688}\right)\right) d\xi$$

Solution to the resulting equation (15) was done numerically on four orders of accuracy by the Runge-Kutta method [12].

The numerical experiments studied how the geometric dimensions of the pore chamber and the performance of the production well (the share of oil in the production, the total oil production, the steam oil ratio) had been changed depending on the reservoir characterization and the injection well regime. The alteration of the following parameters was observed in calculations: initial temperature, temperature drop dT, pore chamber height h, dry steam mass fraction in steam-water mixture φ, thermal conductivity of rocks λ_r, total well flow rate q.

The calculations were carried out on the following basic values of the parameters: $dT = 100°C$; $\lambda_r = \lambda_t = 2W/m°C$; $C_{vr} = 2.5MJ/m^3°C$, $C_{vt} = 0.65C_{vr}$; $L = 2.3MJ/kg$, $m = 0.3$, $\Delta s_o = 0.55$, $\Delta s_w = 0.1$, $\Delta s_m = 0.65$, $h = 15m$, $\varphi = 0.8$, $q = 20m^3/d$, $V_0 = 15m^3$, $t_0 = 10d$, $t_x = 500d$.

3. Results and Discussions

The primary results of the numerical tests are shown in Figures 2-6. For all the tests, the figures (a) show the change in the share of oil in the production with time, the figures (b) show the change in the oil production rate, (c) show the growth of the steam chamber volume, and (d) show the change in the steam-oil ratio.

The figures demonstrate that the development is distinctly intensified, and the steam-oil ratio grows more slowly if dT drops, that is, at a higher initial temperature of the formation (Fig. 2.a-2.d). On the one hand, the result is obvious, but may be understood in a different way. The conclusion is that when the reservoir is heated up, it does not make sense to heat the steam to the temperatures higher than those of SAGD process; the excessive increase in temperature leads only to excess heat losses and premature growth of the steam-oil ratio.

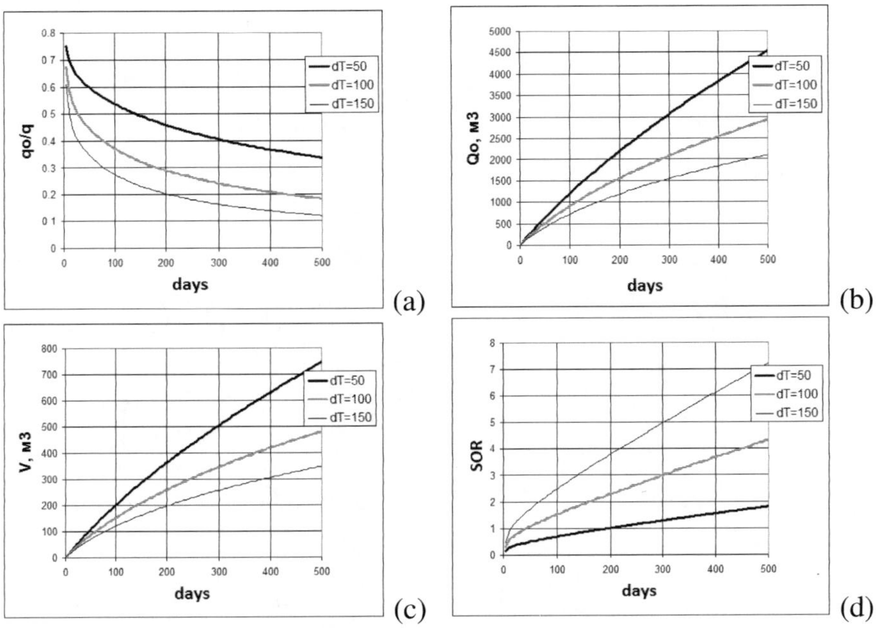

Fig. 2 Changes in measures vs. temperature drops

The increase of the steam chamber in the height results in a proportional increase in oil production and inversely proportional to the reduction in the steam-oil ratio for the same times (Figure 3.a-3.d). This result confirms how the SAGD process is effective for thick reservoirs.

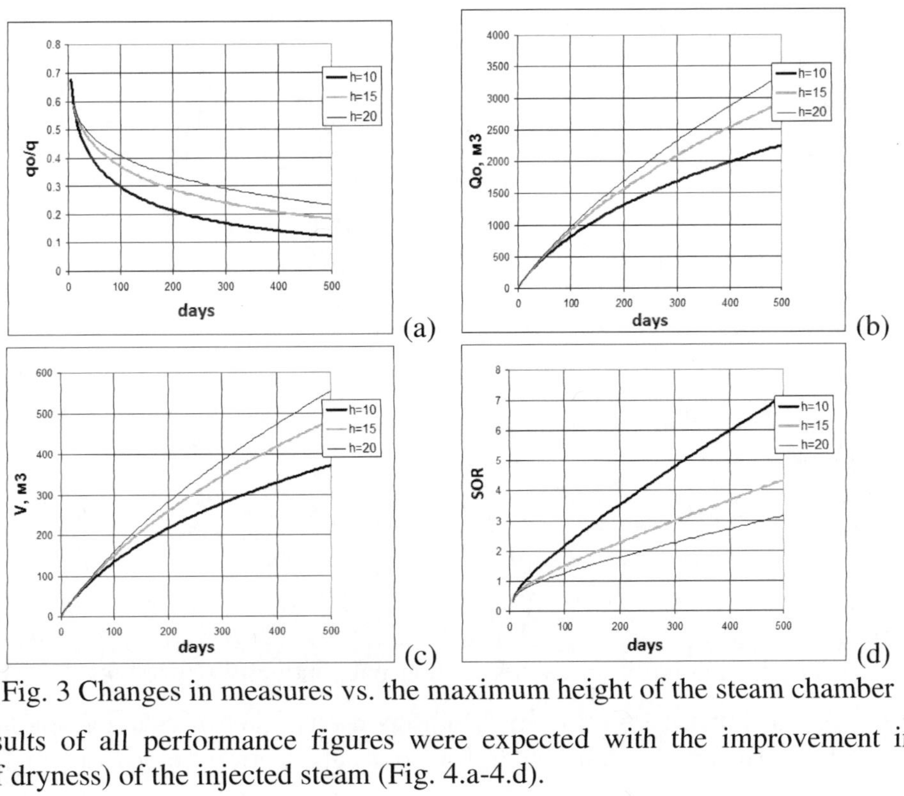

Fig. 3 Changes in measures vs. the maximum height of the steam chamber

Better results of all performance figures were expected with the improvement in quality (degree of dryness) of the injected steam (Fig. 4.a-4.d).

IOP Conf. Series: Earth and Environmental Science **155** (2018) 012016 doi:10.1088/1755-1315/155/1/012016

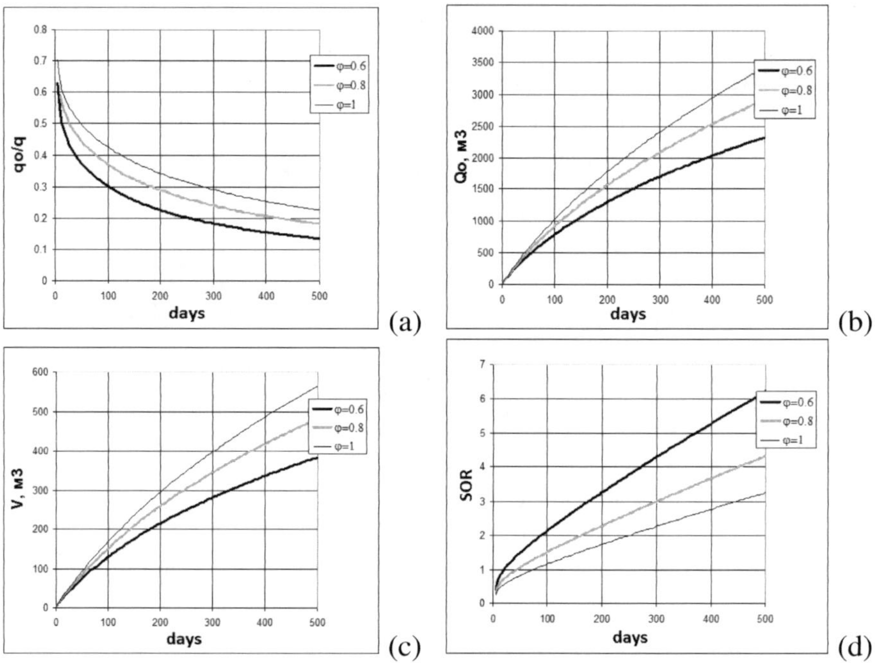

Fig. 4 Changes in measures vs. mass fraction of dry steam

Higher thermal conductivity of rocks increases heat loss from the steam chamber, which, in turn, provokes degradation of the figures. The dependence on λ is quite strong (Figure 5.a-5.d). Therefore, the exact definition of this parameter is particularly challenging.

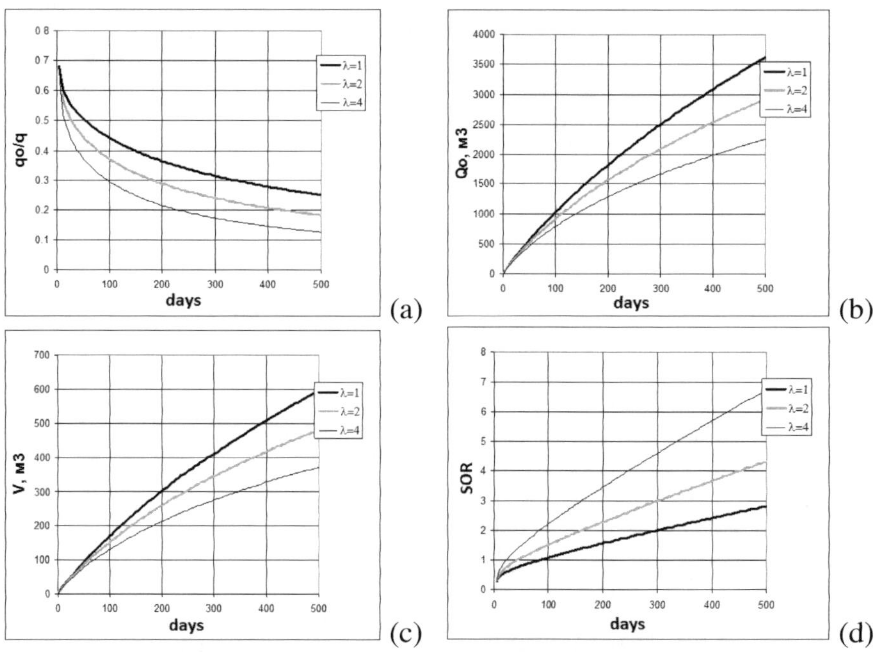

Fig. 5 Changes in measures vs. thermal conductivity of rocks

The change in the total production rate yields a proportional intensification of the production (growth Q_o and V). However, relative figures (water cut, steam-oil ratio) vary slightly (Fig. 6.a-6.d).

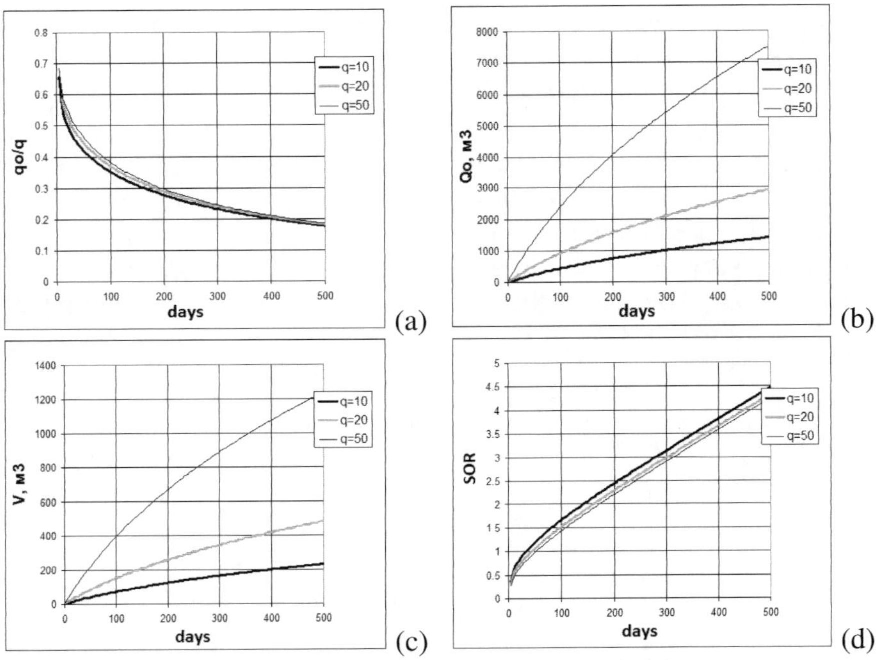

Fig. 6 Changes in measures vs. total production rate

4. Conclusions

The result is an analytical model that is capable to estimate the parameters of the steam chamber near the vertical well. Unlike the numerical calculations in a full-scale formulation, the calculations by the resulting model take an essentially shorter time.

The model can be applied for rapid assessments of the relevance of the SAGD process using in vertical wells as injection ones.

The results of the parametric analysis allow making conclusions about the qualitative dependence of the key development parameters both on the reservoir characterization and on the technological conditions of development.

Acknowledgements

The work was supported by the Ministry of Education and Science of the Russian Federation (project No. 02.G25.31.0170) and by the subsidy allocated to Kazan Federal University as part of the state program for increasing its competitiveness among the world's leading centers of science and education.

References

[1] Butler R M 1996 *Houston: Gulf Publishing Company* 228 p
[2] Heidari M, Pooladi-Darvish M, Azaiez J, Maini B 2009 *Journal of Petroleum Science and Engineering* **68** pp 99–106
[3] Butler R M, Stephens D J 1981 *Journal of Canadian Petroleum Technology* **N2** pp 90-96
[4] Butler R M 1985 *Journal of Canadian Petroleum Technology* **N3** pp 42-51
[5] Klemin D, Pimenov V, Rudenko D 2008 *SPE* 117387
[6] Elliot K T and Kovscek A R 1999 *In, Enghien-les-Bains, France*

[7] Khisamov R S, Morozov P Ye, Khayrullin M Kh, Shamsiyev M N, Abdullin A I 2015 *Neftyanoye hozyaistvo* **N2** pp 62-64
[8] Pimenov V P, Klemin D V, Rudenko D V 2009 *Izvestiya VUZov, Geologiya i razvedka* **N1** pp 49-52
[9] Diwan U K and Kovscek A R 1999 *Technical report, Stanford University, CA, USA*
[10] Pimenov V P, Popov Yu A, Klemin D V, Spasyonnykh M Yu 2009 *Izvestiya VUZov, Geologiya i razvedka* **N4** pp P. 59-62
[11] Polyanin A D, Vyazmin A V, Zhurov A I, Kazenin D A 1998 *Moscow: Factorial* 368p
[12] Bakhvalov N S, Zhidkov N P, Kobelkov G M 2001 *Numerical methods* pp. 363–375

Synthesis and study of the synthetic hydroxyapatite doped with aluminum

M Goldberg[1,*], V Smirnov[1], O Antonova[1], A Konovalov[1], A Fomina[1], V S Komlev[1], S Barinov[1], A Rodionov[2], M Gafurov[2] and S Orlinskii[2]

[1] A.A. Baikov Institute of Metallurgy and Materials Science, Russian Academy of Sciences, 49 Leninsky Pr., Moscow, Russia
[2] Kazan Federal University, 18 Kremlevskaya Str., Kazan, Russia

E-mail: margo.goldberg@yandex.ru (Goldberg Margarita), Tel.: +7-929-651-63-31.

Abstract. Powders of synthetic hydroxyapatite doped with aluminium (Al) ions in concentrations 0 and 20 mol. % were synthesized by the precipitation method from the nitrate solutions and investigated by atomic emission spectrometry with inductively coupled plasma (AES-ICP), X-ray diffraction (XRD), scanning electron microscopy (SEM), gas absorption and conventional electron paramagnetic resonance (EPR). It is shown that for the chosen synthesis route an introduction of Al provokes formation of highly anisotropic phase, leads to the decrease in the crystallinity while no significant changes in the EPR spectra of the radiation-induced defects is observed. The results could be used for understanding the structural transformations with Al doping of the mineralized materials for geological and biomedical applications.

1. Introduction

The element phosphorus is widely distributed in the earth's crust, but because of its great affinity for oxygen is not found in nature in a free or uncombined state. In the form of orthophosphate (PO_4) phosphorus also combines readily with a number of cations, chiefly calcium, to form numerous phosphate minerals. There is known to be more than 150 phosphate minerals but only the calcium phosphates, mainly those of the apatite group, and, to a very limited extent the aluminium (Al) phosphates, are generally found in phosphate deposits of commercial importance [1].

Phosphate minerals containing significant amounts of aluminium (and/or iron) are to be found mainly in tropical or subtropical regions where they are formed by intensive lateritisation of phosphatic bedrock. An understanding of the structural complexity of mineralized materials and their transformations under the influence of the environmental factors such as soil pH, soil hydrology, temperature, etc. is fundamental for exploration into the field of diagenesis. For example, unusual replacement of biogenic apatite in dinosaur teeth by aluminium and iron phosphates as a result of fossil diagenetic processes was observed in Spain [2].

During recent years there have been efforts in developing synthetic calcium phosphate to enhance their biological and mechanical properties for use in biomedical applications that can be mainly achieved by cationic substitutions of calcium ions [3]. Aluminium can serve as one of the substituents but the biological effect of Al remains controversial (both positive and negative functions of the element in human bone metabolism have been reported) while introduction of Al ions can lead to the significant lattice distortions [5]. In our previous work we have demonstrated enhanced solubility of Al in the β-tricalcium phosphate matrix synthesized by precipitation method [6]. In present paper we use the same reaction route to synthesize aluminum substituted hydroxyapatite (Al-HA) powders and to investigate them by different analytical methods to follow the changes caused by aluminum introduction.

2. Materials and Methods

Samples of Al-HA with the chemical formula $Ca_{10-x}Al_{2x/3}(PO_4)_6(OH)_2$ were synthesized with $x = 0$; 0.2 by the precipitation method according to the following reaction using reactants of analytical grade and deionized water:

$$(10\text{-}x)Ca(NO_3)_2+6(NH_4)_2HPO_4+(2x/3)Al(NO_3)_3+8NH_4OH$$
$$\rightarrow Ca_{(10-x)}Al_{2x/3}(PO_4)_6(OH)_2+ 20NH_4NO_3+6H_2O.$$

Calcium and aluminum nitrate solutions were mixed with a diammonium hydrogen phosphate solution in the corresponding ratios. The pH value of the reaction mixture was maintained at a level of 9.0–9.5 by adding aqueous ammonia. Powders were ripened at mother solution during 21 days at temperature of 25°C for full crystallization of precipitate [7]. The obtained powders were filtered, washed and dried at 60°C for 24 h. The powders were heat treated at 300°C for 3 h in air to remove ammonium nitrate, an impurity phase.

For determination of Al content in the precipitates, powders were heat treated at 900 °C for water removing, dissolved in HCl – HNO_3 mixture and analyzed by atomic emission spectrometry with inductively coupled plasma (AES-ICP, HORIBA Jobin Yvon, ULTIMA 2). The powder materials were characterized by the X-ray diffraction (XRD) method (Shimadzu XRD-6000, CuKα radiation) with identification of phase composition according to JCPDS and PCPDFWIN databases. The particle morphology of synthesized powders was assessed by scanning electron microscopy (SEM) (Tescan VEGA II microscope, secondary electron imaging, accelerating voltage of 20kV). The specific surface area (S) of powders was determined by low-temperature nitrogen adsorption measurements (BET, Micromeritics TriStar analyzer).

Continuous wave (CW) EPR spectra of the samples were obtained using table-top Labrador (Ekaterinburg, Russia) and ESP-300 (Bruker) spectrometers. operating at 9.4 – 9.9 GHz (X-band) microwave frequency. PC concentration was estimated at room temperature in the double cavity ER4105DR of ESP-300 by comparing the integrated intensities of the spectra of a test sample and a reference samples (Cu-DETC solution and Mn^{2+} in MgO powder). X-ray irradiation of the synthesized powders was provided by using URS-55 tube (U = 55kV, I = 16mA, W anticathode) at room temperature with the estimated dose of 5 kGy to create stable paramagnetic centers in the nominal pure material.

3. Results and Discussion

Data on Al content in the investigated species are presented in Table 1 which shows that the chemical composition of synthesized materials is close to the predetermined one.

Table 1. Powder composition and surface area for the investigated species.

Ca substitution by Al, mol.%	Stoichiometric formulae	Al theoretical, mol. %	Al theoretical, wt. %	Al measured, wt. %	S, m^2/g
0	$Ca_{10}(PO_4)_6(OH)_2$	0	0	$< 10^{-4}$	91±1
20	$Ca_8Al_{1.33}(PO_4)_6(OH)_2$	20	3.629	3.62 ± 0.04	90±1

According to the XRD analysis, both powders consist of apatite-like phases with the low crystallinity degree (Figure 1). The only detected phase in both samples was hydroxyapatite (JCPDS

#9-432). The shape of triplet between 30.5 ° and 33.5 ° of HA sample became smoother and transformed into the duplet with introduction of Al due to decrease of crystallinity degree of powders in correspondence with ref. [8].

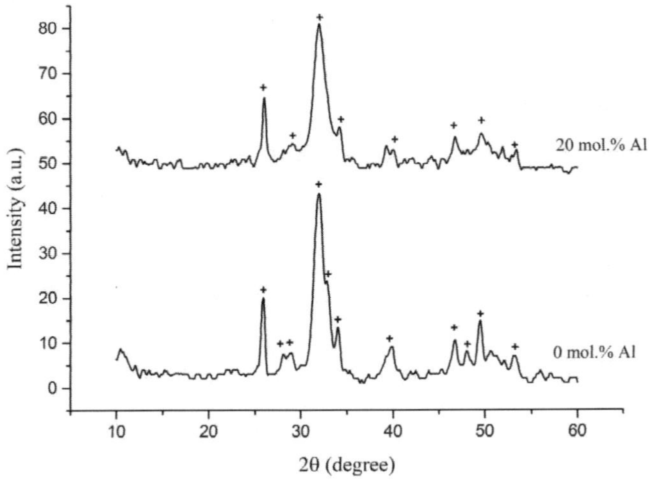

Figure 1. XRD spectra of Al-substituted HA powders. Peaks marked with "+" correspond to HA (JCPDS No: 9 -432).

SEM data demonstrate formation of powders agglomerates for both species. As it can be seen, HA are formed by agglomerates of uniaxial and prismatic particles with average sizes of 50-100 nm (Figure 2). Materials with 20 mol.% of Al are formed by compact agglomerates consisted of 100-250 nm needle-like and prismatic crystals. Thus, doping of HA by Al provokes formation of highly anisotropic phase.

Figure 2. Microphotographs of HA with 0 mol.% (left panel) and 20 mol. % of Al (right panel).

According to the gas absorption data, aluminum introduction does not change the specific area of the synthesized powders that is in the range of 90-91 m^2/g (Table 1).

Pure HA and Al-HA samples are supposed to be EPR silent. Indeed, in the investigated species no EPR signals within the sensitivity limits of our equipment was found that confirms the purity of the initial reagents and proper synthesis treatment [9]. The paramagnetic centers appear after the X-ray irradiation (Figure 3). The EPR patterns are very close to each other and mainly are due to the known in HA synthesized by the wet precipitation technique from the nitrate containing reagents CO_2^- and NO_3^{2-} stable radicals [10]. As we have shown previously in refs [9], introduction of even non-paramagnetic impurities can lead to the significant changes of the relaxation and spectroscopic parameters of the radiation induced defects. From the presented EPR results it follows that the spectroscopic parameters of the obtained radiation-induced defects do not change with the aluminum doping and, therefore, gives an opportunity to follow electron-nuclei interactions with the Al nuclei in the way similar to the previously described for other aluminum containing matrices [12].

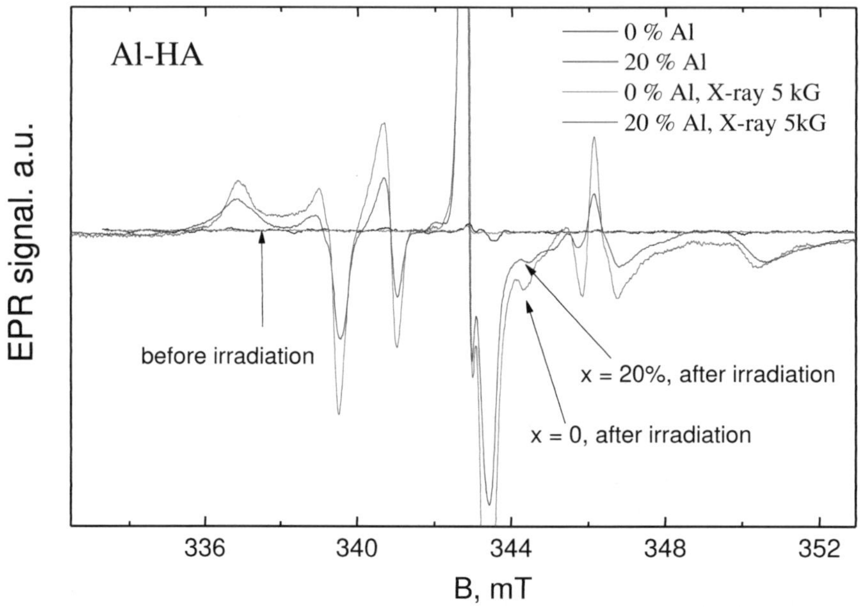

Figure 3. EPR spectra of Al-HA samples before and after X-ray irradiation.

4. Conclusion

Calcium phosphates are compounds of interest to many fields of science, including geology, chemistry, biology and medicine due to their abundance in the nature and presence in the living organism. A huge number of publications were and are devoted to the cationic and anionic substitutions in the hydroxyapatite – they can significantly affect the properties of the material even in very low concentrations. Despite the large number of the studies performed, many important problems related to anionic and cationic substitutions in HA are not thoroughly investigated. Data of different studies contradict one another. The most inconsistent information is associated with the sites of the ions localization in biomineral, synthetic and nanosized samples, influence of dopant onto the biologically relevant properties [9]. We hope that our study is the first step in revealing the role aluminum doping, location of Al ions in the HA lattice and the influence of Al on the local environment in HA structure.

IOP Conf. Series: Earth and Environmental Science **155** (2018) 012017 doi:10.1088/1755-1315/155/1/012017

Acknowledgements
The powders synthesis and XRD, SEM, AES-ICP and gas absorption investigations were performed with financial support of Russian Science Foundation under the grant # 16-13-00123. EPR analysis was conducted with support of state assignment № 007-00129-18-00.

References

[1] Notholt A, Highley D, Slansky M 1979 *Dossiers IV. Phosphate. Raw Materials Research and Development*, (Cedex: Commission of the European Communities) p 234
[2] Bauluz B, Gasca J, Moreno-Azanza M, Canudo J 2014 *Lethaia* **47** 556
[3] Barinov S 2010 *Russ. Chem. Rev.***79** 13
[4] Laskus A, Kolmas J 2017 *Int. J. Mol. Sci.***18** 2542
[5] Wang M, Wang L, Shi C, Sun T, Zeng Y, Zhu Y 2016 *Phys. Chem. Chem. Phys.* **18** 21789
[6] Goldberg M, Smirnov V, Protsenko P, et al. 2017 *Ceram. Int.* **43** 13881
[7] Goldberg M, Smirnov V, Ievlev V, et al. 2012 *Inorg. Mater.* **48** 181
[8] Kolekar T, Thorat N, Yadav H, Magalad V, Shinde M, Bandgar S, Kim J, Agawane G 2016 *Ceram. Int.* **42** 5304
[9] Murzakhanov F, Gabbasov B, Iskhakova K, et al. 2017 *Magn. Reson. Solids* **19** 17207
[10] Gafurov M R, Biktagirov T B, Mamin G V, Shurtakova D V, Klimashina E S, Putlyaev V I and Orlinskii S B 2016 *Phys. Solid State* **58** 469
[11] Biktagirov T, Gafurov M, Mamin G, Klimashina E, Putlayev V, Orlinskii S 2014 *J. Phys. Chem. A* **118** 1519
[12] Herklotz F, Lavrov E, Weber J, Mamin G, Kutin Y, Volodin M, Orlinskii S 2011 *Phys. Status Solidi B* **248** 1532
[13] Mukhambetov I, Lamberov A, Yavkin B, Gafurov M, Mamin G, Orlinskii S 2014 *J. Phys. Chem.C* **118** 14998

IOP Conf. Series: Earth and Environmental Science **155** (2018) 012018 doi:10.1088/1755-1315/155/1/012018

Radiation induced paramagnetic radicals in synthetic octacalcium phosphate

M Gafurov[1,*], B Yavkin[1], G Mamin[1], S Orlinskii[2], I Smirnov[2], A Fedotov[2] and VS Komlev[2]

[1] Kazan Federal University, 18 Kremlevskaya Str., Kazan, Russia
[2] A.A. Baikov Institute of Metallurgy and Materials Science, Russian Academy of Sciences, 49 Leninsky Pr., Moscow, Russia

*E-mail: marat.gafurov@kpfu.ru

Abstract. Created at room temperature by X-ray irradiation at the dose of about 5 kGy radiation induced centres in stable octacalcium phosphate (OCP) obtained by double transformation of α-tricalcium phosphate (α-TCP) are studied by means of the conventional and pulsed electron paramagnetic resonance (EPR). No EPR signals within the sensitivity of the used equipment in the non-irradiated samples were detected. In the irradiated species complex EPR spectrum signals due to H^0 and CO-like radicals appear. The obtained results could be used for the tracing of the mineralization processes from its initiation to the completion of the final product and for the *in situ* identification of the OCP phase.

1. Introduction

Among the non-destructive methods which can give an invaluable information about the content and structure of various materials is electron paramagnetic resonance (EPR) which detects paramagnetic impurities, trapped free radicals or hole centres [1]. EPR is also used for dating minerals and rocks containing paramagnetic defects, cave deposits with defects in calcite, biocarbonate defects in fossils and sediments as shells, corals or mollusks, apatite related defects in bones, tooth enamels and dentine. Human and animals bones and teeth are suitable for dating, for an estimation of absorbed dose of irradiation, for detection of irradiated foodstuffs containing bones or fish-bones (see papers [2-4] and references therein). A number of precursor minerals have been suggested to be important in the biomineral evolution of Earth's mineralogy, including amorphous calcium phosphate – hydroxyapatite (HA), tricalcium phosphate (TCP), octacalcium phosphate (OCP), and brushite [5].

Synthetic and sintered calcium phosphates, first of all hydroxyapatites, irradiated with γ-rays and X-rays or doped with paramagnetic impurities have been intensively studied by EPR and by electron-nuclear resonance (ENDOR) techniques giving a background for interpretation of EPR spectra of natural materials [3]. Several paramagnetic species located at hydroxy or phosphate sites have been identified in synthetic HA: oxygen radical O^-, trapped atomic hydrogen, and hole trapped on OH^- and PO_4^{2-}. Carbonate radicals CO_2^-, CO_3^-, and CO_3^{3-} are often observed as well as additional organic radicals (alanine-like) signals and color centres signals appears. In burnt bones the coal-type C• radicals were identified [2, 3]. In HA synthesized by the solid state phase methods or wet-synthesis technique from the nitrate containing substances stable H^0 and/or NO_3^{2-} radicals, correspondingly, are often detected [5-10].

Even for HA despite of decades of investigation, some problems of EPR spectra separation and unambiguous assignment exist, especially in samples of biogenic origin [11]. Already for the secondly frequently used CaP - tricalcium phosphate amount of the EPR studies is practically negligible [10, 12]. Since recently, octacalcium phosphate as one of the minerals formed by organisms has attracted increasing attention because of good osteoconductivity, biocompatibility, and biodegradability [13-15]. In particular, OCP is considered to be a better scaffold for bone regeneration in comparison to hydroxyapatite or to β-tricalcium phosphate. To the knowledge of the authors, the only OCP related research of the radiation induced paramagnetic centres is paper [16] in which OCP was used as a precursor for HA while EPR in HA was investigated. We present the preliminary results of our study of the X-ray created paramagnetic centres in stable OCP at conventional for EPR frequency of 9 GHz (X-band).

2. Materials and Method

The crystal structure of OCP with the chemical formulae $Ca_8H_2(PO_4)_6 \cdot 5H_2O$ is triclinic with a space group of $P\bar{1}$ (the center of inversion is located at (½,½, ½), the unit cell contains 110 atoms) and exhibits the much longer *a*-axis dimension than the *b*. OCP has the lattice constants $a \approx 19.8$ Å, $b \approx 9.6$ Å, $c \approx 6.87$ Å, $\alpha \approx \beta \approx 90^o$ and $\gamma \approx 109^o$. It resembles the structure of HA, $Ca_{10}(PO_4)_6(OH)_2$ with the corresponding hydroxyapatite constants for $P6_3/m$ hexagonal group $a \approx 18.84$ Å, $a = b \approx 9.42$ Å, $c \approx 6.885$ Å, $\alpha = \beta \approx 90^o$ and $\gamma \approx 120^o$ [17]. In ref. [18] it was first shown that OCP consists of apatite and hydrated layers. It has been suggested that a progressive evolution from OCP to apatite can occur by a topotactic reaction corresponding to the elimination of the hydrated layer.

Synthesis of α-TCP was carried out by precipitation from solutions of calcium nitrate and ammonium hydrogen phosphate:

$$3Ca(NO_3)_2 + 2(NH_4)_2HPO_4 + 2NH_4OH \rightarrow Ca_3(PO_4)_2 + 6NH_4NO_3 + 2H_2O. \quad (1)$$

A solution of ammonium hydrogen phosphate (0.5 mol/l) was gradually added to a solution of calcium nitrate (0.5 mol/l) with constant stirring. The resulting suspension was allowed to age in the air during the 24 hours. The precipitate was then filtered off and dried at 120 °C.

For the production of α-TCP ceramic granules, the technology of impregnating a cellular polymeric template with ceramic slip was used [19], followed by sintering at 1300 °C with quenching. The quenching was carried out by cooling the ceramics at room temperature. After that, the ceramic was crushed and granules of 500-100 μm were sifted.

α-TCP ceramic granules were transformed to OCP according to the route described in ref [20]. Briefly, 10 g of α-TCP ceramic granules were immersed to 1000 ml of 1.5 M sodium acetate aqueous solution and 0.15 ± 0.02 M glutamic amino acid, adjusted with orthophosphoric acid to a pH value of 5.5 ± 0.1. The granules were kept in for 168 h with shaking at 35 °C. The granules were thoroughly washed in distilled water at least five times and dried overnight at 37 °C. Then, the granules were immersed in an aqueous 1.5 M sodium acetate solution with a pH value of 8.7 ± 0.1 for 168 h with shaking at 35 °C.

The phase composition was analyzed by conventional X-ray diffraction (XRD) technique (Shimadzu XRD-6000 (Japan), Ni-filtered CuKα1 target, λ=1.54183 Å). Samples were scanned from 2Θ ranging from 3° to 40° with a step size of 0.02° and a preset time of 5 second.

Continuous wave (CW) EPR spectra of the samples were obtained using table-top Labrador (Ekaterinburg, Russia) and ESP-300 (Bruker) spectrometers. operating at 9.4 – 9.9 GHz (X-band) microwave frequency. PC concentration was estimated at room temperature in the double cavity ER4105DR of ESP-300 by comparing the integrated intensities of the spectra of a test sample and a reference samples (Cu-DETC solution and Mn^{2+} in MgO powder). Pulsed EPR measurements were done exploiting the X-band abilities of Bruker Elexsys 580/680 spectrometer. X-ray irradiation of the synthesized powders was provided by using URS-55 tube (U = 55kV, I = 16mA, W anticathode) at room temperature with the estimated dose of 5 kGy to create stable paramagnetic centers in the nominal pure material.

3. Results and Discussion

The XRD pattern of the product obtained after soaking in standard buffer solution for 168 h is shown in Figure 1. XRD of the investigated samples confirmed OCP phase with following characteristic: (010) main reflection at 2Θ = 4.9° and peaks (2 6 0), (151) and (150), (0 7 0), (0 0 2).

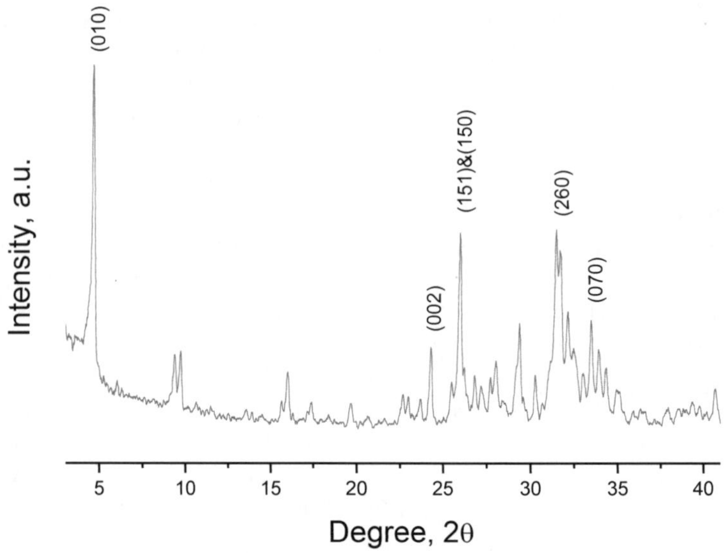

Figure 1. XRD pattern of OCP

Pure, non-substituted OCP (as well as the mentioned HA and TCP) is supposed to be EPR silent. Consequently, EPR might be used for the purity check of CaP materials [10]. As Figure 2 demonstrates, in the investigated samples no EPR signal was detected at room temperature (i.e. concentration of the paramagnetic impurities $C < 10^{15}$ spin/g) while after the irradiation a signal in the central part of the spectrum appears with the total concentration of $C \approx 8(1)\cdot10^{16}$ spin/g.

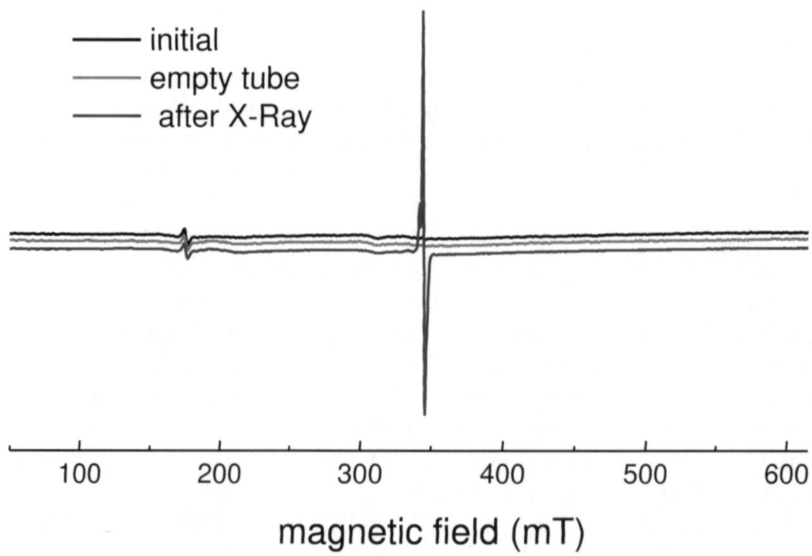

Figure 2. Conventional room-temperature X–band EPR of the initial sample, empty tube (glass holder) and X–irradiated sample of OCP.

The central part of the spectrum (in the magnetic field range of 300-400 mT) for the irradiated species was investigated in details. From Figure 3 it follows that a pair of weak lines with the splitting of about 500 mT due to the H^0 (nuclear spin $I = 1/2$) could be observed. This feature is also characteristic for the HA and TCP materials synthesized by solid state reaction [10]. But the part of the spectrum in the range of the magnetic fields of 340-350 mT resembles those for HA and TCP synthesized by the wet precipitation techniques [9]. The remarkable differences for CaP produced by various methods could be exploited for the fairy simple identification of the synthesizing route of the commercial CaP materials, for example which is often a proprietary information.

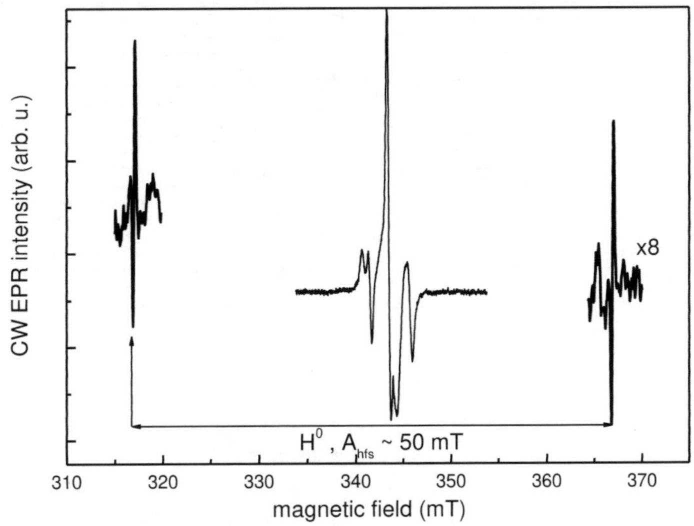

Figure 3. Conventional room-temperature X–band EPR of X–irradiated sample of OCP.

We have studied the behavior the signal in the range of the magnetic fields of 340-350 mT comprehensively. Figure 4 presents the dependence of the spectrum on microwave power. At least two different sorts of paramagnetic centres could be distinguished: while the intensity and amplitude of the broad line in the middle of the spectrum with the linewidth of 1 mT is growing with power (denoted as a paramagnetic centre 1, PC_1), while the second "structural" paramagnetic centre (PC_2) with the with the maximal splitting of 55 mT starts to saturate already at $P_{MW} = 10$ mW. Lineshape of PC_1 resembles those for the carbonate centered stable radicals in HA [3]. To study the nature of PC_1 in details we have measured transverse (T_2) and longitudinal (T_1) relaxation times at T = 100 K by pulsed methods ($\pi/2$-τ-π and π-T-$\pi/2$-τ-π pulse sequence, correspondingly) in the magnetic field marked by the blue dashed line in Figure 4. T_2 curve can be described by only one exponential decay with $T_2 = 1.8$ μs. T_1 curve is presented in Figure 5.

Experimental points on T_1 curve were approximated by one-, two- and three exponents with the corresponding T_1 times(1) $T_1 = 0.5$ ms; (2) $T_1^{(1)} = 0.06$ ms, $T_1^{(2)} = 0.7$ ms; (3) $T_1^{(1)} = 0.02$ ms, $T_1^{(2)} = 0.2$ ms, $T_1^{(3)} = 1.2$ ms. As it follows from Figure 5, three-exponential description is the best one. Assuming that the centre with the longest T_1 belongs to PC_2 (cf. Figure 4 and discussion below), there are two types of the fast relaxing centres in OCP.

EPR pattern of the slowly relaxing PC_2 resembles that in HA. In the irradiated HA samples EPR is mainly due to the stable NO_3^{2-} ions preferably substituting one of the PO_4^{3-} position in the HA structure (substitution of B-type). Their powder EPR spectra can described by the spin-Hamiltonian of the axial symmetry

$$H = g_{\|}\beta B_z S_z + g_{\perp}\beta\left(B_x S_x + B_y S_y\right) + A_{\|} S_z I_z + A_{\perp}\left(S_x I_x + S_y I_y\right), \tag{2}$$

where $g_{\|}$ and $g_{\perp}$ are the main components of the g tensor, $A_{\|}$ and $A_{\perp}$ are the main components of the hyperfine tensor, B_i, S_i and I_i are the projections of the external magnetic field strength, $S = ½$ and

$I = 1$ are electronic and nuclear spin correspondingly onto the $i = \{x, y, z\}$ coordinate axis, β is a Bohr magneton. The HA spectra could be simulated properly if we supposed a Gaussian (continuous) distribution of $A_{\|}$ with a deviation of 0.40 mT around the mean value of 6.65(40) mT and a discrete distribution of $A_{\perp}$ around 3.37(5) mT, while the components of g-factors for all of the obtained radical's modifications are the same ($g_{\|} = 2.0011(1)$, $g_{\perp} = 2.0052(1)$) [6].

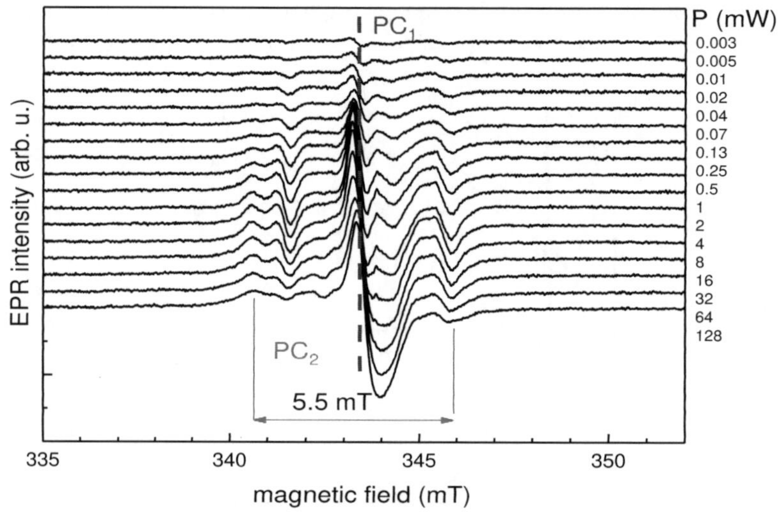

Figure 4. Conventional room-temperature X–band EPR of the X–irradiated sample of OCP at different microwave power levels. Lines with different microwave saturation behavior are marked.

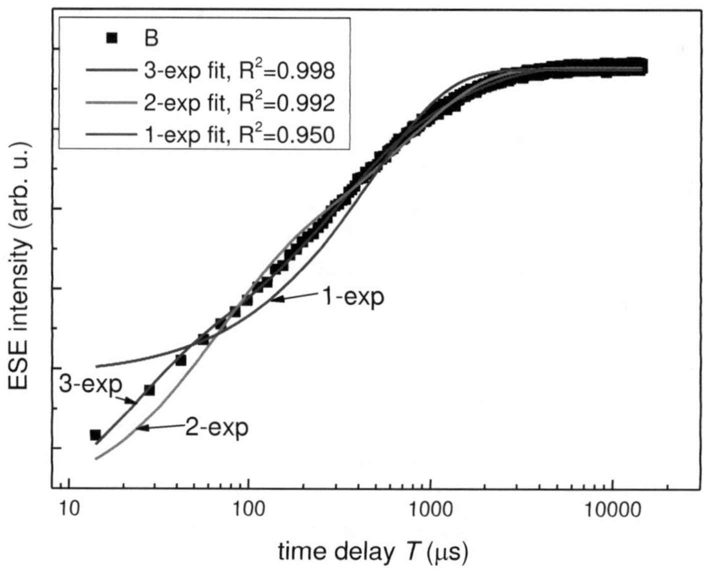

Figure 5. T_1 curve (filled squares) with the corresponding approximations by one-, two- and three exponents.

In our paper [10] we have found that EPR spectrum of the wet synthesized TCP after radiation is also mainly due to the NO_3^{2-} with the same as for HA value of $A_{\|}$ (A_1 is 2 times $A_{\|}$ of 6.65 mT) but the distribution of $A_{\perp}$ (A_2, A_3) is larger than that for the HA. We have suggested that it could be due to the three different positions for PO_4 group substitution in TCP structure as well as due to the various schemes of the charge compensation. As it follows from the presented results, in OCP we have to suggest $A_{||} = 5.5/2 = 2.7$ mT. Such small value forces to suggest the presence of other nitrogen containing radicals such as NO_2, for example [6], or other paramagnetic centres [3].

4. Conclusion

Modern EPR techniques offer various opportunities for the rapid testing and comprehensive analysis of the condensed matter. The capabilities of even conventional EPR approaches for investigation and analysis of calcium phosphate materials are still not fully exploited and described in details. We hope that our paper would serve as a first step in understanding the nature and structure of various paramagnetic centres in octacalcium phosphates with the purpose of their usage as sensitive intrinsic probes to follow the changes under the doping or surface treatment.

Acknowledgments

The work was funded by the subsidy of the Russian Government to support the Program of competitive Growth of Kazan Federal University among World's Leading Academic Centers, within the state task № 007-00129-18-00 and Program of Fundamental Research of the Presidium of the Russian Academy of Sciences 37 P.

References

[1] Khasanova N, Gabdrakhmanov D, Kayukova G, et al. 2017 *Magn. Reson. Solids* 19, 17102
[2] Wencka M, Hoffmann S and Hercman H 2005 *Acta Physica Polonica A* **108** 331
[3] Fattibene P and Callens F 2010 *Applied Radiation and Isotopes* **68** 2033
[4] Skinner A 2011 *Radiation Measurements* **46** 749
[5] Hazen RM et al. 2008 *American Mineralogist* **93** 1693
[6] Gafurov M et al. 2014 *Appl. Magn. Reson.* **45** 1189
[7] Gafurov M et al. 2014 *JETP Lett.* **99** 196
[8] Biktagirov T et al. 2014 *J. Phys. Chem. A* **118** 1519
[9] Gafurov M et al. 2015 *Phys. Chem. Chem. Phys.* **17** 20331
[10] Murzakhanov F et al. 2017 *Magn. Reson. Solids* **19** 17207
[11] Chelyshev Y et al. 2016 *BioMed Research International* **2016** 3706280
[12] Fadeeva I et al. 2017 *BioNanoScience* **7** 434
[13] Suzuki O 2010 Acta Biomaterialia **6** 3379
[14] Shen D et al. 2017 *Bio-Medical Materials and Engineering* **28** 9
[15] Shi H, He F and Ye J 2016 *J. Mater. Chem. B* **4** 1712
[16] Moens P et al. 1994 *Appl. Magn. Reson.* **6** 121
[17] Matsunaga K 2010 *J. Am. Ceram. Soc.* **93** 1
[18] Brown W 1962 *Nature* **196** 1048
[19] Fedotov A, Bakunova N, Komlev V, Barinov S 2011 *Doklady Chemistry* **439** 233
[20] Komlev V et al. 2014 *ACS Appl. Mater. Interfaces* **6** 16610

IOP Conf. Series: Earth and Environmental Science **155** (2018) 012019 doi:10.1088/1755-1315/155/1/012019

Effect of lithological heterogeneity of bitumen sandstones on SAGD reservoir development

E A Korolev[1], S A Usmanov[1], D S Nikolaev[1] and R R Gabdelvaliyeva[1]
[1]Institute of Geology and Petroleum Technologies, Kazan Federal University, Kazan, Russia

E-mail: sausmanov@gmail.com, nikolaevcd@gmail.com

Abstract. The article describes the heavy oil field developed by the SAGD method. While development planning all the heterogeneity of the reservoir is must be taken into account. The objective of this work is to identify the distribution of lithological heterogeneities and their influence on oil production. For this reason, the studies of core samples were conducted and the heterogeneity was identified. Then properties and approximate geometry of lithological objects were studied. Also the effect of the heterogeneity on the heat propagation and production of fluid were analyzed. In the end, recommendations were made for the study of such heterogeneities on other deposits with similar geology

1. Introduction

Russia is rich in heavy oil and natural bitumen reserves, which are in a stepwise process of commercial development [1]. One of the production project sites of hydrocarbons difficult to recover are the Ufimian bitumen fields in the southeast of Tatarstan. The heavy oil and bitumen reservoirs are currently being developed, which pay zones occur at a shallow depth (50-200 m). One of the primary methods of heavy hydrocarbon recovery is the steam-assisted gravity drainage (SAGD). At sites, the steam simulation consists of injecting steam through a pair of horizontal wells so creating a steam chamber and warming up the reservoir layer to reduce the viscosity of the oil. The main difficulty to control this process is the uneven distribution of heat between the wells and the leakage of steam out of the chamber [2,3].

To understand the possible geological reasons for such negative phenomena, detailed lithological and mineralogical tests of wells were carried out on one of the heavy oil reservoir in the Republic of Tatarstan. The uplift chosen as a test target is one of the bar features forming the ridges stretched along the coast of the Permian paleosea of the Ufimian age [4,5,6] (Fig. 1). Considering the morphology and arrangement of ridges in the paleorelief, the bars and barrier islands are assumed to have been formed on shallow terraces where the underwater ledges of the seabed developed. The area seems to rise slightly for a short period during the formation of the Sheshma upper unit thus resulted in the stabilization of the shoreline, which ensured a long-term processing of detrital material under the conditions of active shallow-water hydrodynamics. As a result, clay siltstones and sandstones underlying the bar structures were replaced by well-sorted fine-grained sands building up underwater shafts and above-water islands. Underwater wave breaking and bottom migration of sediments contributed to a local increase in the thickness of sandstones within bar features, to improve lithological homogeneity and grading of their detrital material. This produced the accumulative forms of the coastal-offshore relief. The subsequent transgression of the Permian paleosea changed the conditions of sedimentation, which resulted in the deposition of mostly clay material in the "lingual clays" unit. The thick "lingual clay" drapes overlapping the Sheshma sand unit became a kind of regional screen allowing oil deposition in sand bars during the upward migration of water-oil fluids. Eventually, oil deposits transformed into heavy oils and bitumen due to oxidation processes.

2. Methodology

The core analysis showed that the hydrocarbon deposits within the studied uplift are confined to well graded sandstones composing the bar body. The underlying clay-calcareous sandstones and siltstones are only pigmented with oil. The bitumen deposit is around 20 m thick in the dome of the structure.

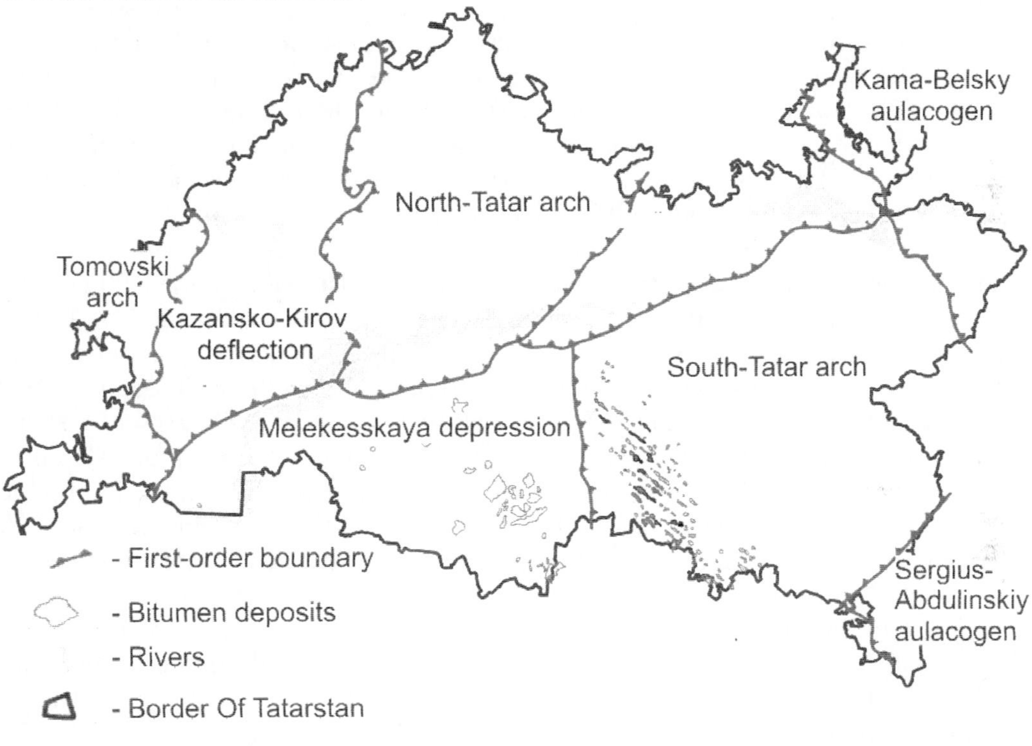

Figure 1. Map of heavy oil and bitumen field distribution in the Sheshma horizon bar deposits in Tatarstan.

The bitumen reservoir is lithologically represented by greywacke sandstones [7]. The optical microscopic tests show that the detrital component accounts 70-85% and cemented material accounts 15-30% in the rock. The allothigenous assemblages are dominated by rounded fragments of siliceous and effusive rocks while the subrounded grains of quartz, angular grains of albite and microcline, chlorite flakes, rare rounded carbonate fragments are less presented. The fractional composition of sandstones is dominated by grains 0.1-0.25 mm in size, the silt and pelitic grains are in subordinate amounts. The fragments of rocks and minerals are cemented by argillo-calcareous and bituminous cement. The mineral cement (10-15%) of pore-lining and vein-cluster types having a micro-fine-grained texture, and the bituminous cement (85-90%) is a pore-lining cement. A relatively small amount of argillo-calcareous cement causes the incoherent texture of sandstones within the bitumen reservoir. If the bitumen substance is removed, the open porosity of the rock is 15-30 %. At the same time, it tends to reduce to 10-12 % as it approaches the bottom of the productive portion of the bitumen reservoir. The X-ray computed tomography shows the communicating pores evenly distributed in the rock volume are 30-80 μm in effective size [8].

Studying the mechanical properties of the sandstones of the examined heavy oil reservoir revealed rocks featuring poor structural bonds. Almost throughout the entire bitumen reservoir, their compressive strength is 1.25-2.75 MPa, the tension strength is 0.05-0.11 MPa, the bond strength between mineral grains is 0.21-0.45 MPa, the Protodyakonov scale of hardness accounts up to 0.3 MPa (Category VII).

Similar lithologic-petrographic features of bitumen sandstones show a good heat carrier intake capacity of the terrigenous reservoir. Theoretically, considering the differences in reservoir pressure (0.4-0.5 MPa) and injection pressure (2.0-2.5 MPa), the injected water vapor should be relatively well and uniformly distributed between the two wells and up-dip producing a cylindrical high-temperature steam chamber. However, the well logging data show that the propagation of the heat carrier is difficult in some cases [2].

When handling the core samples, it was noted that a consolidated interval (10-50 cm) occur among the medium- and semi-consolidated bitumen-saturated sandstones (Fig. 2). At certain hypsometric levels, they can be traced in several wells drilled in different parts of the uplift, that is, they exhibit an area distribution pattern, but not traceable in the well logging data due to the insufficient resolution of the latter [9]. Optical-microscopic and XRD tests have shown that all such layers are formed by tightly cemented sandstones. And, if the allothigenic component is almost identical to those of soft sandstones in composition and ratio of mineral fragments in the rock, the composition and texture of cement differ significantly. Secondary basal-pore carbonate cement occurs in tight layers, which is also a cement that fills secondary pores (completely fills all pores). Unlike the micro-fine-grained cement of soft sandstones, the cement within the consolidated sandstone interval is uniform, fine and medium-grained in texture. Some intervals are dominated by calcite in the mineral composition of cement, while in the others – by dolomite. The porosity of tight sandstone does not exceed 6 %. These are predominantly subcapillary pores.

Figure 2. Productive interval core of well 171 with consolidated sandstone interval (red line).

The results of the geological core description and the core studies were then used to build the surface of the consolidated sandstone interval top. The wells of this interval were selected for the further analysis (Fig. 3) followed by the A-B line using the saturation cube (Fig. 4).

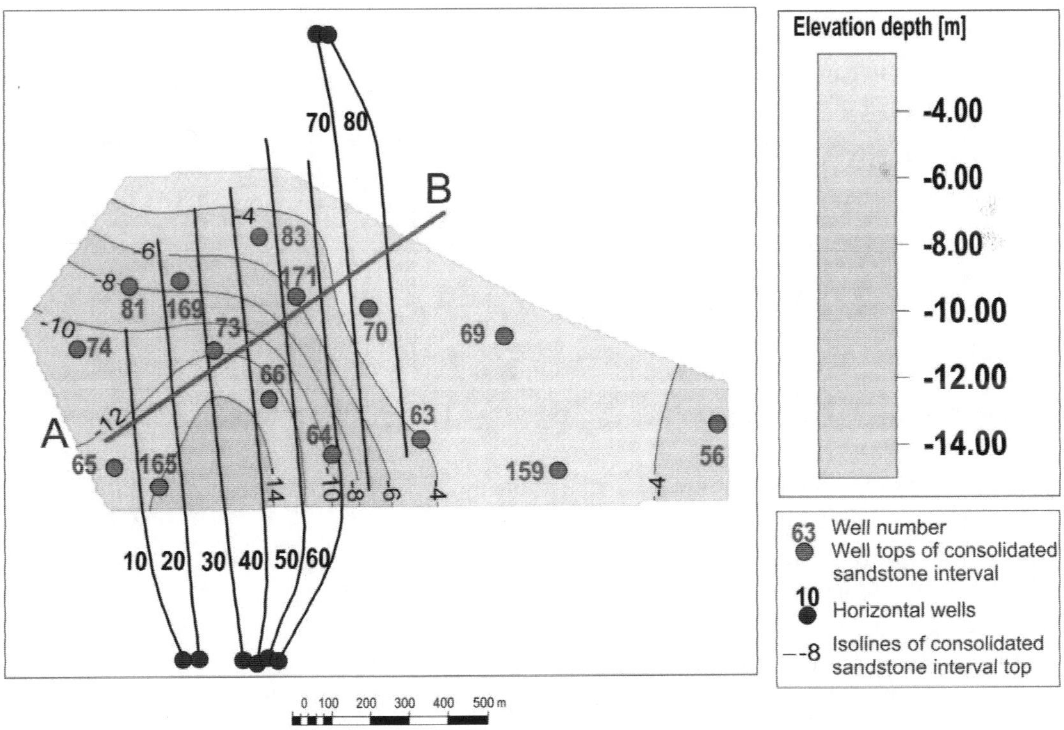

Figure 3. Top of consolidated sandstone interval map.

IOP Conf. Series: Earth and Environmental Science **155** (2018) 012019 doi:10.1088/1755-1315/155/1/012019

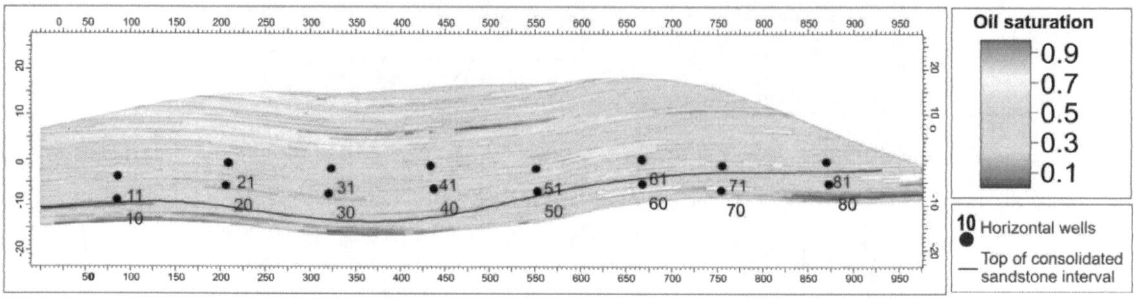

Figure 4. Cross-section of oil saturation (A-B line).

The preliminary paleostructural analysis has revealed the tendency of spreading the consolidated sandstone interval according to the bottom of the Sheshma bitum bearing interval. The interval occurs below the producing well with regard to the pairs of horizontal wells 10-11 – 50-51. In the area of pairs of horizontal wells 60-61 – 80-81, the interval crosses the space between the producing and injection wells.

The following technological parameters of the development were analyzed: cumulative production of liquid and oil, steam injection volumes, oil/steam ratio (O/S), water cut (WC) (Table 1). The production and injection data are normalized on a scale of 1-10, where 10 is the maximum production/injection in the present range.

Table 1. Development parameters of the horizontal well pairs considered

WELL PAIR	OIL PRODUCED	WATER PRODUCED	LIQUIDS PRODUCED	STEAM INJECTED	WC	O/S	MEAN TEMP, °C	INJECTION PERIOD, DAY
10-11	6.6	8.4	8.3	8.5	0.95	0.07	70	242
20-21	3.2	7.1	6.8	7.6	0.97	0.02	41	349
30-31	4.1	5.1	4.9	5.5	0.97	0.04	83	170
40-41	10.0	10.0	10.0	6.8	0.94	0.14	96	303
50-51	4.6	4.0	4.0	4.6	0.93	0.09	83	180
60-61	1.2	6.7	6.3	7.4	0.99	0.02	51	303
70-71	0.4	6.1	5.7	9.3	0.99	0.01	39	297
80-81	1.1	9.9	9.4	10.0	0.99	0.01	33	377

With a glance to the relatively smooth distribution of flow parameters calculated by well logging data that were captured between these two wells, one of the factors affecting well performance figures is the presence and location of the above-mentioned consolidated sandstone interval in the vicinity of the wells.

Following the analysis of the accumulated development figures, preliminary conclusions were made about the possible effect of the consolidated sandstone interval on the development parameters given in Table 1.

Where the interval occurs between the pair of wells being a production and an injection well, the effect of decreasing the effective vertical permeability of the reservoir as a decrease in the oil production rates, is visible.

3. Results and Discussions

Wells Nos. 60 and 70 separated from the respective injection wells by the consolidated sandstone interval, demonstrate low efficiency and considerable water cut. In addition, despite the significant volumes of injected steam, these wells are less heated in the perforation zone against the well pairs 10-11, 30-31, 40-41 and 50-51. The possible reason of this effect could be the vertical heterogeneity of the section due to the presence of a low-permeability interval that impedes the propagation of heated oil and condensed fluid from the upper part of the formation where the steam chamber [10] is formed to the production well. The temperature drop can be caused by the displacement of the drainage area of the production well to the lower interval of the reservoir saturated with formation water having a low temperature. In this case, the heat propagation is guaranteed mostly by conductive heat transfer and the hindered communication of the heated upper interval and the production well in this area can cause a reduced performance efficiency of the well pairs 60-61 and 70-71. It should be noted that the considered effects supposedly take place in the pair of wells 80-81, however, one cannot ignore the reduced oil-saturated thickness of deposits near the wells 80-81 (Fig. 4).

The well performance analysis (Table 1) show that the steam-oil ratio is higher for well pairs 40-41, 50-51 and 10-11. This parameter can be used to assess the efficiency of a particular pair of wells. Among these high-performance wells, the pair of wells 40-41 is characterized by a significant amount of injected steam and an extensive oil-saturated thickness, which is likely to cause good performance of the well. These parameters for the other two high-performance pairs of wells (50-51 and 10-11) can be described as moderate, but production wells in these pairs are not much higher than the consolidated sandstone interval (Fig. 4). It is assumed that one of the effects that may occur due to the presence of a consolidated sandstone interval below the horizontal production well may be the limitation of water inflow into the production well from the underlying water-saturated units. Moreover, the effects of downward oil flow below the production well are under consideration. These effects should have a positive effect on well pairs 50-51 and 10-11.

4. Conclusions

Since the consolidated sandstone intervals are hard to distinguish in the bitumen reservoir using well logging techniques and surface surveys due to their small thicknesses, it is proposed to consider the propagation of the heterogeneous zones based on core tests when planning the trajectories of paired horizontal wells. Taking into account a similar geological history of the heavy oil and bitumen reservoir development in Tatarstan, similar effects can occur in other reservoirs recovered by injecting steam into the reservoir. The allocation of heterogeneity intervals, i.e. represented by consolidated sandstone, similar to that considered in this paper, should be performed at the stage of the general study of the reservoir and the results should be counted in order to construct more detailed geological models.

The work was supported by the Ministry of Education and Science of the Russian Federation (project No. 02.G25.31.0170) and by the subsidy allocated to Kazan Federal University as part of the state program for increasing its competitiveness among the world's leading centers of science and education.

References

[1] Danilova, Ye.A. *Heavy Oil in Russia* // The Chemical Journal. – 2008. – No. 12. – p. 34-37.

[2] Sudakov, V. *Technology of Integrated Monitoring of Steam Chamber Evolution During the Oil Production from the Shallow Deposits of Super-Viscous Oil* / Sudakov V., Nurgaliev D., Khasanov D., Stepanov A., Khamidullina G., Kosarev V., Galukhin A., Usmanov S., Zaripov A., Amerkhanov M. // Paper SPE-182000-MS presented at the SPE Russian Petroleum Technology Conference and Exhibition, 24-26 October 2016, Moscow, Russia. http://dx.doi.org/10.2118/182000-MS

[3] Makarevich, V.N., Iskritskaya, N.I., Bogoslovsky, S.A. *Resource Potential of Heavy Oil Reservoirs in the European Russia* // Neftegazovaya Geologiya. Teoriya i praktika. – 2012. – T. 7, No. 5. – p. 27–32.

[4] B.V, Burov. *Geology of Tatarstan. Stratigraphy and Tectonics* – M.: GEOS, 2003. – p. 402.

[5] Petrov, G.A. Lithological *and Facial Analysis of the Upper Permian Bitumen Reservoirs Related to the Assessment of Bitumen Resources Within Tatarstan*: Author's abstract. Thesis Work by Candidate of Geological and Mineralogical Sciences. – Kazan, 2000. – p. 27.

[6] Khasanov R.R., Mullakaev A.I., Dusmanov E.N. *The structure of sandstones in productive horizons of the Permian bituminous deposits of Tatarstan (Russia)* // Uchenye Zapiski Kazanskogo Universiteta. Seriya Estestvennye Nauki. – 2017. – Vol. 159, № 1. – p. 164–173.

[7] Mullakaev A.I., Delev A.N., Usmanov S.A., Sudakov V.A., Khasanov R.R. *Tectonic causes of uneven cementation zones distribution in the bituminous sandstones productive part of the Sheshmian horizon of the South Tatar arch* // Neftyanoe Khozyaystvo - Oil Industry. – 2018. – №2. – p. 23-25.

[8] Zakirov, T.R., Galeev, A.A., Korolev, E.A., Nuriyev, I.S., Statsenko, E.O. *Estimation of sandstone reservoir properties using X-ray CT studies in Ashalchinskoye oil field* // Neftyanoe Khozyaystvo - Oil Industry. – 2015. – № 8. – p. 36-40.

[9]. Nikolaev, D.S. *Studying the Geological Features of the Primary Pay Zones of Heavy Oil Reservoirs in Tatarstan* // West Siberian Petroleum Congress: a collection of scientific papers of the XI International Scientific and Technical Congress of the Student Department of the Petroleum Engineer Society / Editor-in-chief S.I. Grachyov; TIU. – Tyumen: TIU, 2017. – p.169. ISBN 978-5-9961-1461-0

[10] Butler R.M. Horizontal Wells or the Recovery of Oil, Gas and Bitumen // Petroleum Society Monograph No. 2, 1994.

Laboratory Tests of Bitumen Samples Elasticity

E R Ziganshin [1], S A Usmanov[1], D I Khasanov [1], G S Khamidullina [1]
[1]Institute of Geology and Petroleum Technologies, Kazan Federal University, Kazan, Russia

E-mail: eduard-ziganshin@mail.ru, sausmanov@gmail.com

Abstract. This paper is devoted to the study of the elastic and acoustic properties of bitumen core samples. The travel velocities of the ultrasonic P- and S-waves were determined under in-situ simulation conditions. The resulting data were then used to calculate dynamic Young's modulus and Poisson's ratio. The authors studied the correlation between the elasticity and the permeability and porosity. In addition, the tests looked into how the acoustic properties had changed with temperature rise.

1. Introduction

Bitumen deposits are an essential type of hydrocarbon reservoirs that contribute significantly to the world reserves. Heavy oil reservoirs contain reserves of more than 6 tln. barrels which is three times the total world reserves of conventional oil and gas [5]. The capacity of this resource seems very attractive, but petrophysical characterization of these reservoirs requires deeper understanding for the effective development of them.

Unlike conventional hydrocarbons, bitumens are more viscous. This results in difficulties to apply traditional technologies to develop heavy oils [7]. For the enhancement of production, various techniques have been proposed to reduce the viscosity of bitumen, for example, by injecting chemical solvents or heating up the formation [6, 8]. There are a lot of challenges that are new or poorly understood to be solved at the stage of bitumen reservoir management. In particular, optimization of the well map for production and injection wells, control of coolant propagating within the reservoir. A lot of tests have been done to explore such deposits [1]. However, this scope of an inquiry is still burning and requires more tests.
This paper is designed to studying the acoustic and elastic geomechanical properties of bitumen sandstones under in-situ simulation conditions, as well as to understanding of how these parameters change as the temperature rises. In addition, the correlation between acoustic parameters and the permeability and porosity was studied. The tests can significantly improve the accuracy of the interpretation of shallow seismic survey data (updated velocity model). Correct setting of such correlations as input data will make it possible to identify subtle field changes based on surface tests, which will ensure an increase in the resolution of seismic surveys to monitor the development of shallow reservoirs of heavy oils.

2. Methodology

The paper describes experimental testing on core samples aimed at measuring their acoustic properties and elastic dynamic parameters. The experiments were performed on the PIK-UZ-UEP test set-up (Figure 1).

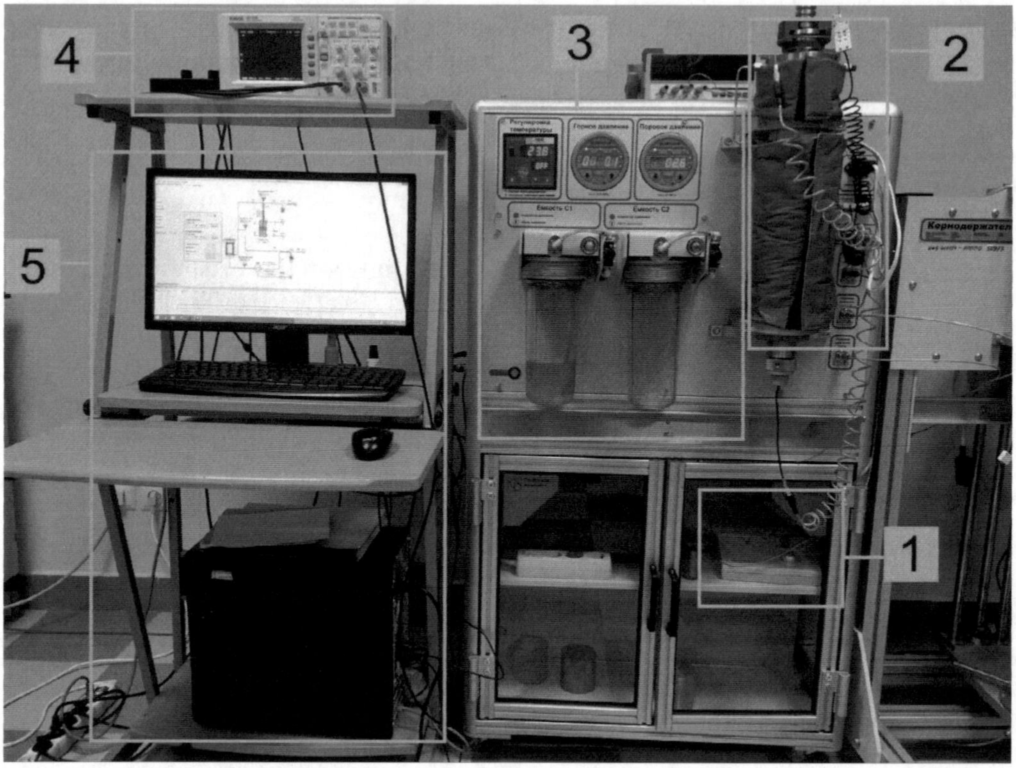

Figure 1. The PIK-UZ-UEP test unit for measuring the elasticity under in-situ simulation conditions: 1 – signal source; 2 – core holder; 3 – control unit for creating reservoir conditions; 4 – the Rigol 1102E oscilloscope; 5 – personal computer with software.

The test unit consists of two ultrasonic sensors, a signal source, an oscilloscope, an RLC meter and relays. The source generates 1 MHz signal. The signal goes to one of the sensors, which excites the pulses of S- and P-waves. The waves pass through a core sample held by a core holder and are detected by the second ultrasonic sensor. Then the received signal goes to the oscilloscope. The oscilloscope is connected to the PC for analysis and recording of the received signal. The source gives three output signals: S-wave, P-wave, and a signal of source synchronization with oscilloscope. When measuring the elasticity of core samples, it is possible to graphically display the S- and P-waves, measure the wave travel time, and capture wave travel velocities, as well as to acquire Young's modulus and Poisson's ratio using the software to control the operation of the test unit. This requires knowing, among others, the length, the diameter, mass and density of the sample. A cylindrical core sample 30 mm in diameter and 30 mm in length is placed in a rubber cup in the core holder, where reservoir conditions are reproduced (squeezing pressure similar to the rock pressure, pore pressure similar to the reservoir pressure, as well as the temperature similar to the reservoir pressure).

Acoustic properties are measured in the following sequence:

1. The sample is inserted into the core holder and clamped by two clamping plungers;
2. The sample squeezing pressure is raised to 5 MPa and the temperature goes up to 25 °C inside the core holder; in the case of poor visibility of the first arrivals, the pressure can rise to 15 MPa;

IOP Conf. Series: Earth and Environmental Science **155** (2018) 012020 doi:10.1088/1755-1315/155/1/012020

3. The oscilloscope connection is done, and the settings of the signal amplifiers are picked to better determine the first arrivals;
4. The operator selects the time of the first arrival and enters its value into the program to calculate the P- and S-wave velocities, Young's modulus and Poisson's ratio;
5. The heating of the sample goes on, and measurements take place every 10 degrees, starting at 30 °C and ending at 90 °C.

3. Results and Discussions

The lab test results of open porosity and density by the sinker method were used to compare the resulting data with the porosity and permeability.

28 samples of the rocks of the sand formation taken from the Ufimian Sheshma horizon, which is the main oil bearing stratum within the Lower Karmalka superviscous oil reservoir of the Cheremshanskoye field, were tested. The lithological analysis identifies 4 types of samples (Table 1):

a) Gray calcareous clay with a shell-like fracture. Non-bitumen samples;
b) Hard gray medium-grained sandstone. Non-bitumen samples;
c) Brown calcareous fine-grained, moderately cemented, uniformly bitumen-saturated sandstone;
d) Clayed gray flag-like limestone with multiple brachiopod shell fragments. Non-bitumen samples.

Table 1. Types of rock core samples by lithological description.

a)	b)	c)	d)

The laboratory tests under reservoir conditions resulted in correlations obtained and studied that allowed to estimate the interrelation between the porosity and permeability and the elasticity of rocks. When comparing the P-wave velocity with the open porosity, it was established that a good correlation is observed in samples of group C (bitumen sandstone) and D (non-bitumen limestone). The ultrasonic P-wave velocity versus porosity correlation shown in Figure 2 can be described by a linear function with a high degree of reliability (R^2 = 0.8282):

$$V_p = -82.805 \times \varphi + 5514.2, \qquad (1)$$

where φ = porosity, %; Vp = P-wave velocity, m/s.

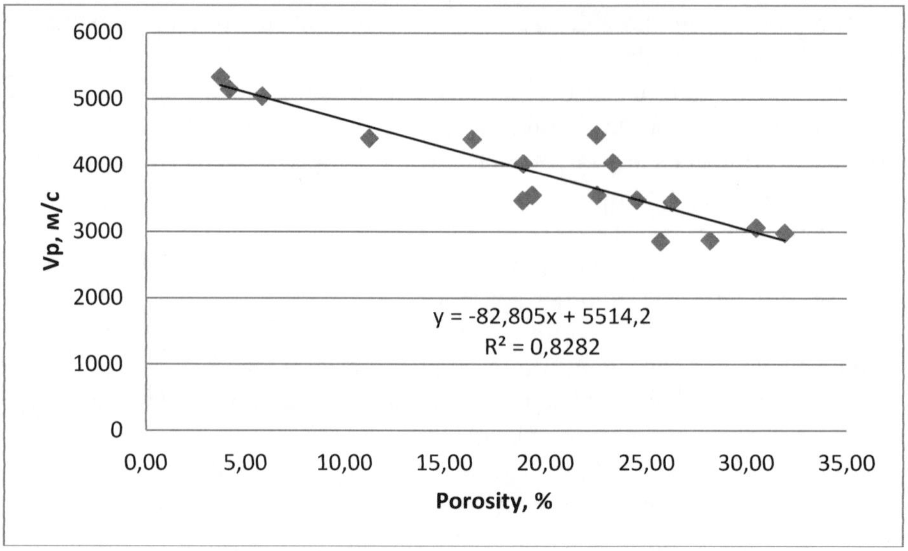

Figure 2. P-wave velocity vs. porosity.

This laboratory method of study acoustic properties can be recommended as an express method for estimating open porosity. Determination of the ultrasonic wave travel velocity under reservoir conditions should not take more than 10 minute for one sample.

These velocities and elastic moduli were calculated for a sample that was heated from 25 to 90 degrees. The measurements were taken every 10 degrees. The resulting figures were then used to look into the dependence of the P- and S-wave velocities, Young's modulus and Poisson's ratio on the temperature. For most of the samples, the velocities and elastic moduli decreased uniformly. In samples containing bitumen (type C, Table 1), a greater decrease in velocity was observed unlike the non-bitumen samples (type A, B, D, Table 1). In bitumen-saturated samples, the P-wave velocity decreased by an average of 6.4 %, and in non-bitumen samples by 4.9 %. For S-wave velocities, in bitumen-saturated samples – 6.9 %, in non-bitumen samples – 4.8%. The same is true for Young's modulus: 12.5 % and 8.4 %, respectively, for bitumen-saturated and non-bitumen samples.

The acoustic properties of samples as the temperature rises can be described by a linear equation of the form Vp = A*T+B, where Vp and T are the P-wave velocity and the temperature, respectively; the factor A describes the rate of change of the acoustic properties. Figure 3 shows the correlation of the factor A (rate of change of acoustic properties) and the porosity. This correlation is described by the linear equation:

$$A = 0.0616 \times \varphi - 5.6118 \quad (2),$$

where A is the factor characterizing the rate of change of temperature, φ is the open porosity, %. The degree of reliability is rather high (R2 = 0.8612); therefore, this correlation can also be used to estimate open porosity.

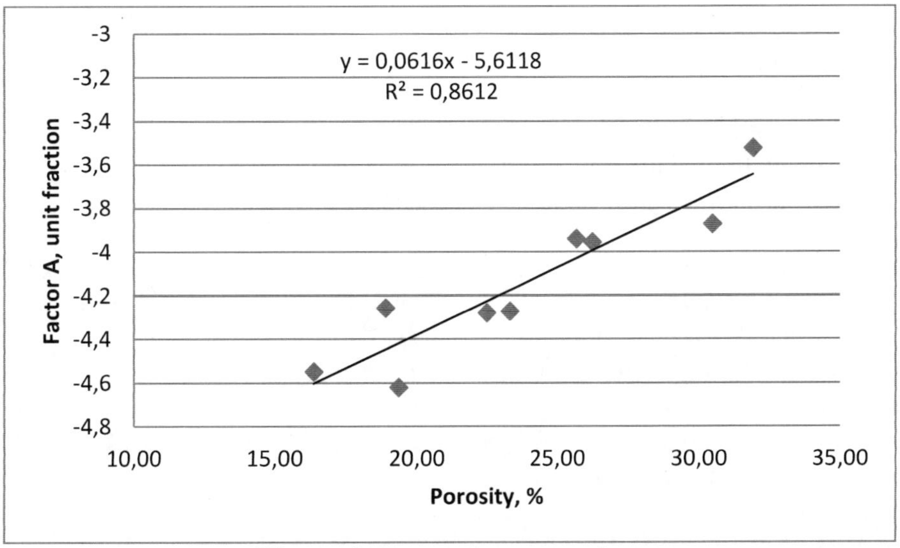

Figure 3. Factor A vs. porosity.

The fact that the samples behave differently when heated arouses much interest. Moreover, the groups identified by the change in velocity are not compared with the groups determined from the point of lithology. A possible explanation may be the pattern of how the bitumen situated in the rock.

According to the previous survey [3], the distribution of bitumen in the rock at low temperatures can be divided into three main types:

1.The sandstone is saturated with water, and the bitumen is sealed off from the pore walls, so the bitumen becomes part of the saturating liquid. Should this happen, the bitumen can mainly change the pore pressure, or its effect on acoustic velocities is similar to that of liquid.

2.The rock is heavily saturated with bitumen, and the bitumen forms a matrix to which sandstone grains are attached without interacting with each other.

3.The rock is less saturated with bitumen, and therefore bitumen accumulates only on grain contacts. Sands are cemented with bitumen, which can significantly increase the through-sample wave velocity.

The type of bitumen sandstone determines the pattern of the ultrasonic wave velocity [2]. The velocity is assumed to decrease linearly for sandstones as the bitumen of type (1) propagates across the rock. Samples behave non-linearly that are most likely related to the propagation type of bitumen (2) and (3). Also, the velocity jump can be caused by the specific dependence of fluid compressibility on temperature. According to Avchyan, "the presence of liquid in pores can cause a total effect of increasing velocity as the temperature rises" if the medium is poorly cemented and saturated with fluid at a small pore pressure [4]. The tests show that there is a need for a more detailed studying the pores of bitumen samples. The next test stage will be computer tomography of these samples to determine the propagation of bitumen and pore space.

4. Conclusions

The laboratory tests of bitumen sandstones reveal the following results:

- The P- and S-wave velocities, as well as the porosity and permeability of bitumen sandstones samples were acquired;
- The correlation of the ultrasonic P-wave velocity and the open porosity was analyzed to an acoustic study practice as an express method for estimating the porosity factor;

- The behavior of the P-wave velocity with increasing temperature was studied, four groups were distinguished according to the change in the velocity values;
- A linear function between the rate of change in acoustic properties and the values of open porosity were determined;
- An attempt was made to explain the pattern of velocity-temperature curves.

References

[1] Eastwood J 1993 *Temperature-dependent propagation of P- and S-waves in Cold Lake oil sands: comparison of theory and experiment Geophysics 58 (6)* pp. 863–872

[2] Han D-H, Liu J and Batzle M 2006 *Acoustic property of heavy oil-measured data: 76th Annual International Meeting, SEG, Expanded Abstracts* pp. 1903–1907

[3] Han X and Junxin G 2016 *Rock physics modeling of acoustic velocities for heavy oil sand. Journal of Petroleum Science and Engineering Vol 145 September 2016* pp. 436-443

[4] Avchyan G M, Matveenko A A and Stefankevich Z B Petrophysics of sedimentary rocks under deep conditions M: Nedra 1979 p. 224

[5] Muslimov R H, Vojtovich E D, Badamshin Je Z, Lebedev N P, Smelkov V M, Uspenskij B V Accommodation and development of natural resources of Tatarstan Geology of oil and gas №2 1995 P. 7-9

[6] Hisamov R S, Gatijatullin N S, Shargorodskij I E, Vojtovich E D, Vojtovich S E Geology and development of natural bitumen deposits of the Republic of Tatarstan p. 295

[7] Hisamov R S, Sultanov A S, Abdulmazitov R G and Zaripov A T Geological and technological features of the development of deposits of high viscosity and super viscous oil 2010 p. 335

[8] Sudakov V, Nurgaliev D, Khasanov D, Stepanov A, Khamidullina G, Kosarev V, Galukhin A, Usmanov S, Zaripov A and Amerkhanov M *Technology of Integrated Monitoring of Steam Chamber Evolution During the Oil Production from the Shallow Deposits of Super-Viscous Oil* Paper SPE-182000-MS presented at the SPE Russian Petroleum Technology Conference and Exhibition 2016 Moscow Russia pp 24-26

Modeling of cobalt-based catalyst use during CSS for low-temperature heavy oil upgrading

R Kadyrov[1], S Sitnov[1] B Gareev[1], G Batalin[1]

[1]Institute of Geology and Petroleum Technologies, Kazan Federal University, Kazan, Russia

E-mail: Rail7777@gmail.com

Abstract. One of the methods, which is actively used on deposits of heavy oils of the Upper Kungurian (Ufimian) sandstones of the Republic of Tatarstan, is cyclic steam simulation (CSS). This method consists of 3 stages: injection, soaking, and production. Steam is injected into a well at a temperature of 300 to 340° C for a period of weeks to months. Then, the well is allowed to sit for days to weeks to allow heat to soak into the formation. Finally, the hot oil is pumped out of the well for a period of weeks or months. Once the production rate falls off, the well is put through another cycle. The injection of the catalyst solution before the injection of steam opens the possibility for upgrading the heavy oil in the process of aquathermolysis directly in the reservoir.

In this paper, the possibility of using a catalyst precursor based on cobalt for upgrading the hydrocarbons of this field in the process of their extraction is represented. SARA analysis on oil saturated sandstones shows an increase in the proportion of saturated hydrocarbons by 11.1% due to the hydrogenation of aromatic hydrocarbons and their derivatives, the content of resins and asphaltenes are remained practically unchanged. A new method for estimating the adsorption of a catalyst based on taking into account the change in the concentration of the base metal before and after simulation of catalyst injection in the thermobaric conditions of the reservoir is proposed. During the study of catalyst adsorption in the rock, when simulating the CSS process, it is found that almost 28% of the cobalt, which is the main element of the catalyst precursor, is retained in the rock.

Keywords: ThEOR, CSS, catalyst, cobalt, SARA, adsorption, heavy oil.

1. Introduction

With the depletion of light and medium oil reserves, heavy oils and natural bitumen become an important source to meet the growing demand for fuel and petrochemicals. Today we know a lot of extraction technology for heavy oil and natural bitumen, which are practically proved to be effective. One of them is the Cyclic Steam Simulation (CSS), that isaccidentally discovered by Shell in Venezuela, and which is actively developed in the oil fields of the Republic of Tatarstan. Including catalytic nanoscale systems in this technology, it leads to refinement of heavy oils due to the aquathermolysis process [1-4].

Field testing (5-7 days) in the Liaohe deposit (China) a water-soluble catalyst salt VO^{2+}, Ni^{2+}, Fe^{3+}, sequential injection of 400 m^3 of water vapor, the catalyst solution and re-injection of 1600 m^3 of steam, oil production increases (in one cycle) by 30-60%, reduction in the viscosity of oil, which is important for oil production and pumping [5]. In another variant, an oil-soluble catalyst using molybdenum as the active metal is used in the same field. The catalyst solution along the steam is introduced into the tank, after which the deposit is closed for 7-10 days. After the action of the catalyst, the viscosity decreases by 78.2%, the content of oxygen, sulfur and nitrogen in the oil decreases. The authors of these studies note slight differences between laboratory tests and field tests in the oil field [6].

In this work Co^{2+} based catalyst precursor, which is synthesized at the Kazan Federal University, is researched. It refers to a class of oil soluble metal compounds, namely the salts of organic acids -naphthenates, tallates [7, 8]. Cobalt carboxylate produces an exchange reaction of the inorganic metal salt with the sodium salt of DTM:

$$C_{17}H_{33}COOH + NaOH \rightarrow C_{17}H_{33}COONa + H_2O \quad (1)$$

Carboxylate of sodium interacts with the cobalt salt when heated:

$$2C_{17}H_{33}COONa + CoSO_4 \rightarrow (C_{17}H_{33}COO)_2Co + Na_2SO_4 \quad (2)$$

It is assumed that the catalyst solution will be pumped in one of the cycles of the CSS. Initially, the layer will be heated, after that the catalyst will be introduced into the reservoir and will spread under the influence of vapor pressure. Then, the catalyst must be adsorbed and hydrocracking must begin at the temperature 178 °C.

The effectiveness of a catalyst based on cobalt in the aspect of reducing the proportion of asphalt-resinous oil compounds in the Ashalchinskoye field extracted using SAGD technology was shown in [2]; as a result of catalytic aquathermolysis (180 °C), the viscosity of the converted oil decreased by 31.1%.

Object of the research is native bitumen in oil saturated sandstone from the Upper Kungurian (Ufimian) stage, that is sampled from the well no. 160 (Nijne-Karamalskoye oil field, the depth interval is 165-169 m).

2. Methodology

Laboratory modeling of the aquathermolysis process is carried out by using a high-pressure reactor (300 ml volume) manufactured by Par Instruments, USA. The model system is a

IOP Conf. Series: Earth and Environmental Science **155** (2018) 012021 doi:10.1088/1755-1315/155/1/012021

compressed sample of a bituminous sandstone of cylindrical shape with a diameter and length of 30 mm with the addition of 10 g water. This system is subjected to a temperature effect (180 ° C) under catalytic conditions. The precursor of the cobalt-based catalyst is introduced in an amount of 2.2% by weight as a solution in a hydrogen donor, based on the mass content of bitumen in the native rock, which is 5.42 wt%. The operating pressure is maintained at 0.22 MPa. The experiment durationis 6 hours after the process has entered the regime.

Separation by the SARA method is carried out by taking into account the methodical recommendations of ASTM D4124-09 and GOST 32269-2013 in several stages. This method is based on the separation of bitumen into four groups of compounds: saturated hydrocarbons, aromatic compounds, resins and asphaltenes (SARA) according to their solubility and polarity [9].

Initially, asphaltenes precipitation is carried out according to the Golde cold method [10], with appropriate preparation of maltenes, including the remaining three groups (saturated hydrocarbons, aromatics and resins) that are separated by using a glass chromatography column by successively eluting with aliphatic and aromatic hydrocarbons from the adsorbent dehydrated neutral aluminum.

To research the adsorption features of the Co^{2+} basedcatalyst laboratory testing under the conditions, which are close to the production field that is made. The core of natural bituminous sandstone of d=30 mm and h=30 mm from Nijne-Karamalskoye heavy oil field of Tatarstan is saturated under the vacuum by solution of the catalyst. After that the core is heated with water up to 180 °C. Then the core is installed to the filtration camera and 200 ml of water is filtered through the sample at temperature 50 °C, cell pressure 7 MPa and pressure difference 1 kPa (reservoir conditions in oil field) (fig 1). After that the core is heated with water up to 180° C. Then the core is installed to the filtration camera and 200 ml of water is filtered through the sample at temperature 50 ^{o}C, cell pressure 7 MPa and pressure difference 1 kPa (reservoir conditions in oil field). The cobalt concentration in tested sandstone is analyzed by using Neptune Plus high resolution multicollector ICP-MS system.

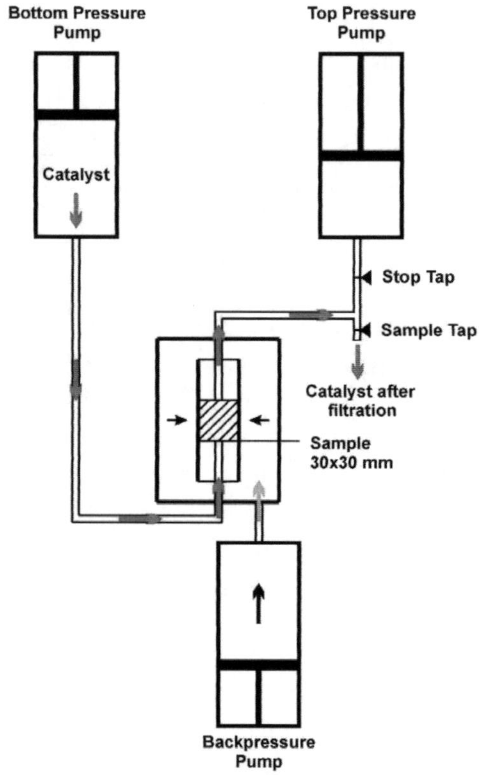

Figure 1. Scheme of the filtration experiment.A sample is installed to the holder and itis pressed up by the water from backpressure pump. Then the filtration of catalyst solutionstarts because of the pressure difference in the bottom and top pumps.

3. Results and Discussions

3.1. SARA group composition

The results of the determination of the composition group of the test samples by the SARA method are given in Figure 2.

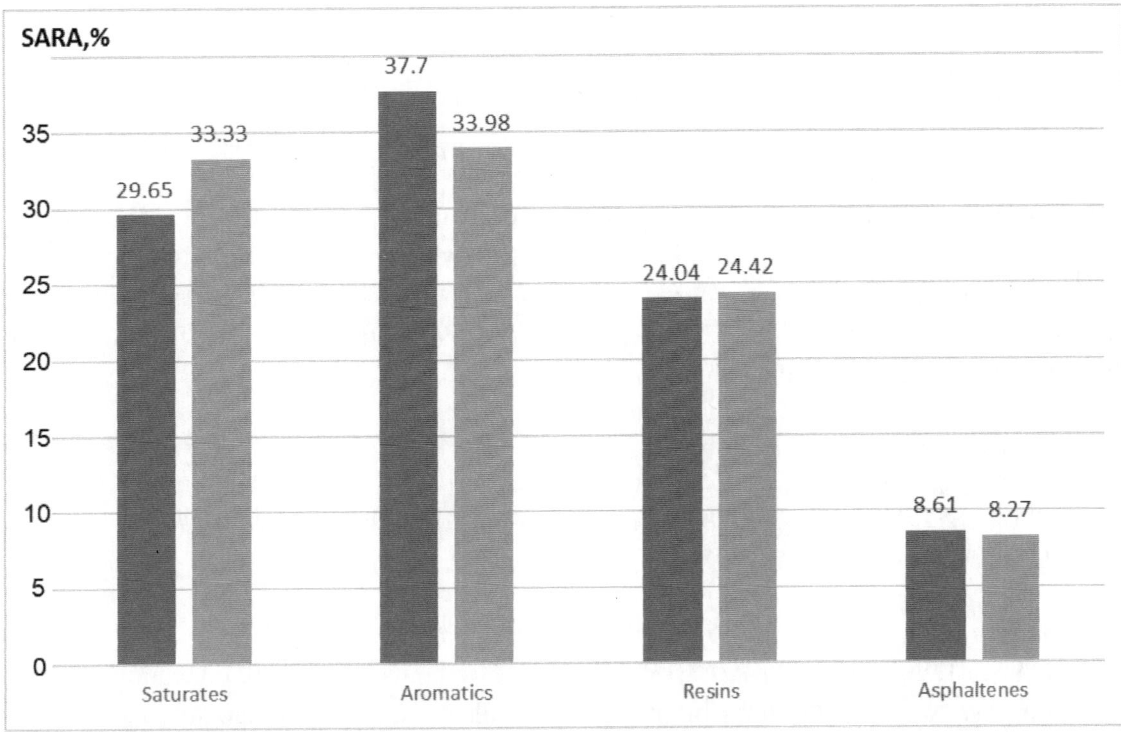

Figure 2. Distribution of SARA composition group in the initial bitumen (white) and after catalytic aquathermolysis (black).

As a result of studies, an increase in the proportion of saturated hydrocarbons is observed, which may occur due to the hydrogenation of aromatic hydrocarbons and their derivatives at the expense of elevated temperature and the presence of a catalyst in a hydrogen donor solution. However, the content of high-molecular components of oil (resins and asphaltenes) does not practically change as a result of catalytic aquathermolysis, apparently because of the short time of exposure to the sample under study.

3.2. Catalyst adsorption

The received data from this testing by this moment show that the biggest part of heavy oil saturated in solvent. The filtration of water through the bitumen sample, after removing catalyst solvent, does not give a big inflow of hydrocarbons. About 28% of catalyst is adsorbed in the sandstone sample, predominantly by clay minerals [11].

Table 1. Cobalt concentrations in the sample

No.	Sample	Cobalt concentration, ppm	Cobalt concentration, %
1	Natural heavy oil saturated sandstone	11.2318	2.5
2	Original catalyst solution	431.9742	97.5
3	Sandstone sample after filtration experiment	123.6636	27.9

4. Conclusions

Thus, the effect of the catalyst precursor on the basis of cobalt (II) is tested for the sandstones of the Nijne-Karamalskoye heavy oil deposits of the Republic of Tatarstan. SARA analysis shows an increase in the proportion of saturated hydrocarbons by 11.1 % due to the hydrogenation of aromatic hydrocarbons and their derivatives. At the same time, the content of resins and asphaltens remains practically unchanged, which may be due to a short exposure time to the sample under study. During the study of catalyst adsorption in the rock, when simulating the CSS process, it is found that almost 28% of the cobalt, which is the main element of the catalyst, is retained in the rock.

Acknowledgements

The work is performed according to the Russian Government Program of Competitive Growth of Kazan Federal University.

References

[1] Khalikova D.A., Petrov S.M., BashkirzevaN.Yu. Review of perspective technologies for processing heavy oils and natural bitumens // Vestnik of Kazan Technologic University. – 2013. – N.3, V.16. – P. 217-221

[2]Sitnov S.A. , Petrovnina M.S., Feoktistov, D.A., Isakov, D.R., Nurgaliev D.K., Amerkhanov, M.I. Intensification of thermal steam methods of production of heavy oil using a catalyst based on cobalt // Neftyanoe Khozyaystvo - Oil Industry. – 2016. – V.2016 (№11). – pp. 106-108

[3]Sitnov S.A., Feoktistov D.A., Kayukova G.P., Pronin N.V., Nosova F.F., Chemodanov A.E. Catalytic intensification of in-situ conversion of high-viscosity oil in thermal steam extraction methods // International Journal of Pharmacy and Technology. –2016. – V.8(3). – pp. 14884-14892

[4]Sitnov S.A., Feoktistov D.A., Petrovnina M.S., Isakov D.R., Darishchev V.I., Akhmadeishin I.A. Structural changes of heavy oil in the composition of the sandstone in a catalytic and non-catalytic aquathermolysis //International Journal of Pharmacy and Technology . –2016. – V.8(3). – pp. 15074-15080

[5] Fan H., Liu Y. Downhole Catalyst Upgrades Heavy Oil // Oil & Gas Journal. — 2002. — V. 100. — N 11. — P. 60–62.

[6] Wen S., Zhao Y., Liu Y. et al. A Study on Catalytic Aquathermolysis of Heavy Crude Oil During Steam Stimulation // 2007 SPE International Symposium on Oilfield Chemistry, 28 February – 2 March 2007, Houston.

[7] S.K.Maity, J.Ancheyta, G.Marroquın. Catalytic Aquathermolysis Used for Viscosity Reduction of Heavy Crude Oils: A Review // Energy & Fuels. — 2010. — V. 24. — P. 2809–2816.

IOP Conf. Series: Earth and Environmental Science **155** (2018) 012021 doi:10.1088/1755-1315/155/1/012021

[8] Weissman J. G. Review of Processes for Downhole Catalytic Upgrading of Heavy Crude Oil // Fuel Processing Technology. — 1997. — V. 50. — P. 199–213.
[9]Feoktistov D.A., Sitnov S.A., Vahin A.V., Petrovnina M.S., Kayukova G.P., Nurgaliev D.K. The description of heavy crude oils and the products of their catalytic conversion according to SARA-analysis data // International Journal of Applied Engineering Research. – 2015. – V. 10. – pp. 45007 – 45014
[10] Bogomolov AI, Abryutina N.N. Modern methods of oil research // L.: Nedra, 1984.- 431 p.
[11] Ivanov VV (1995) Ecological geochemistry of elements, Reference in 6 volumes, v. 4, General d-elements.Nedra, Moscow.

IOP Conf. Series: Earth and Environmental Science **155** (2018) 012022 doi:10.1088/1755-1315/155/1/012022

The first data on the porous space structure of the Domanik shales as a potential object for EOR applying

R Kadyrov[1], E Statsenko[1]

[1]Institute of Geology and Petroleum Technologies, Kazan Federal University, Kazan, Russia

E-mail: Rail7777@gmail.com

Abstract. The resources of shale oil, contained in the organic matter of the wood deposits, can be considered as a source of profitable production of hydrocarbons, when modern EOR technologies are used. As a result of the primary studies of the pore space structure, it is revealed that two types of porous space are prevailing in the studied samples of the Domanik oil shales. The most prevailing is intrakerogen porosity with pore volumes of 5×10^{-8} 1×10^{-6} mm^3. The volumetric reconstruction of the structure of this pore space shows that the voids are confined directly to micro lenses of organic matter. The second type of the found void is represented by leaching cracks. It is characteristic of more carbonate varieties of the Dominik oil shale with spotted structure. It is the oil shale intervals with such cracks that are of greatest interest to the EOR, since they consist of a large area with smaller pores and through which pressurization and spread of various agents are possible to occur in order to increase the oil recovery.

Keywords: Domanik, porous space, shale, source rock, shale oil, unconventional resources, microCT, X-ray computed tomography, EOR.

1. Introduction

Under the conditions of the traditional energy resources depletion, shale and high-viscosity bituminous oils are becoming increasingly important in the world economy. The resources of oil shale - "immature" oil contained in the organic matter of the wood deposits can be considered as a source of profitable production of hydrocarbons, when modern technologies are used. Synthetic oil from the oil shale contains more organic matter than, for example, tar sands of Canada, developed on an industrial scale [1].

It was believed that the Domanik deposits were only a proven high-yielding oil-bearing stratum that generated hydrocarbons for most of the deposits in the overlying carbonate structures of the Volga-Ural and Timan-Pechora basins. At the same time, the Domanik horizon itself contains a considerable amount of hydrocarbons, which have not yet emigrated.

This gives the right to consider the Dominik stratum as a single unstructured hydrocarbon deposit, from which part of the hydrocarbons has left and migrated into structural traps, and most of them have remained and they are the underexplored hydrocarbon resources [2].

Domanik deposits of the Volga-Ural Basin are a high-carbon thin-bedded formation that is capable of both producing hydrocarbons with its own oil and gas reservoirs and concentrating them in separate strata and zones that serve as a reservoir. About 10 oil fields are discovered here, the extraction of which is not carried out. Its inflows are from 15 to 120 tons per day. The technically recoverable oil resources of Domanik are 898 million tons [2, 3].

In the last decades, the applications of horizontal well technology have been widely facilitated by the surging of unconventional reservoirs. At a low drawdown, a horizontal well can have a larger productivity in comparison with vertical wells. The major advantage of horizontal well technology is to enhance the contact area with the formation. Hydraulic fracturing has received the great recognition for one of the most effective techniques for improving the productivity of unconventional reservoirs. Combination of horizontal well with multi-stage hydraulic fracturing and development of different EOR methods, such as cyclic steam simulations (CSS) and cyclic gas stimulation (CGS) techniques can promote the booming of development of shale oil reservoirs [4].

Shale oil and shale gas are present in organic rich rocks, which may act both as the source and the cap rock and can be mixed with clay minerals. Organic materials preserved in rocks have come from the decay of plants and animals. The sediments can retain much of their original organic material in anoxic environment where there is less consumers to consume organic matter. As more material accumulates and underlying ooze becomes compacted, the sediments are buried deeper and subjected to heat and pressure. The organic material slowly and partially is made and transformed into kerogen, depending on the time and temperature at which these materials are made. The transformation of living organisms through diagenesis into kerogen, kerogen will break down to form hydrocarbons through a chemical process called catagenesis [4].

Shales are fine-grained rocks, which are formed from the compaction of silt and clay sized particles. Shales are differentiated from other claystones and mudstones in that they are laminated, finely layered and fissile that mean, they can be broken along their laminations [5]. Shales can have complex mixtures of minerals, and the relative connections of the constituents that have the potential to make or break a potential resource play [4]. This is the reason why research of porous structure of shales is important for unconventional resources recovery of the Domanik formation.

2. Samples and methodology

Two core samples from the Domanik formation of the Republic of Tatarstan were researched. First one was gray carbonate shale, the second – a typical black finely layered and fissile shale. For this study X-ray computed microtomography (microCT) was used. For microCT cylinder samples with 2,5 mm in diameter and 2,5 mm in height were made and 3 ηm resolution of tomography was reached. MicroCT was made by micro and nanofocus X-ray control system for computed tomography General Electric V|TOME|X S 240. Virtual models of the porous structure and measurements performed by Volume Graphics 2.1 and Avizo Fire 8.0 software.

3. Results and Discussions

The obtained results give the first information about the structure of the rock and the porous space of the Domanik oil shale. On the density X-ray orthogonal sections of the sample 1, a spotted texture is distinguished, the dark parts of which correspond to the areas, which are the most enriched in kerogen (Figure 1). Light gray parts are confined to areas with a high content of minerals - in this case mainly calcite. The brightest white points are inclusions of pyrite formed by bacterial sulfate reduction under reducing conditions. In the researched sample, a diagonal crack formed due to leaching is clearly observed. It is noted that it goes round the light-gray mineralized carbonate areas.

Figure 1. MicroCT orthogonal slices of the sample 1. In the picture the spotted texture is observed, where dark zones are interoperated as kerogen rich, light gray zones – carbonate rich, white – pyrite crystals. On the top right part of the picture a leaching crack is represented.

The sample 2 is distinct in a thin-bedded texture (Fig. 2). The dark zones are enriched with kerogen, light grey – quartz. As distinct from the first sample, the amount of pyrite and pores is significantly low.

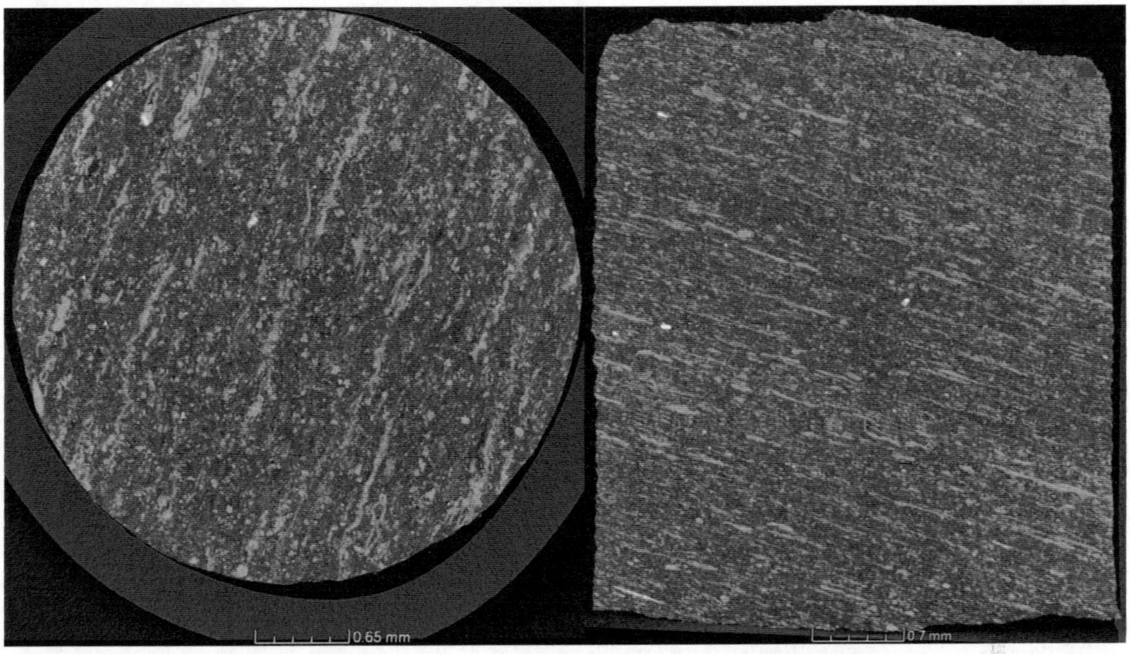

Figure 2. MicroCT orthogonal slices of the sample 2. Thin bedded texture is observed: dark zones correspond to kerogen rich areas, light gray zones – quartz rich, white – for pyrite crystals.

The information on porosity and pore size distribution in the studied samples is given in the table and in the Fig. 3. The porosity value coefficient for the shales is about 1%. In both samples, an equal pore distribution is observed in the volume range from 5×10^{-8} 1×10^{-6} mm^3. We associate such ranges of pore volumes with intra kerogenous porosity. This is indirectly confirmed by the correspondence of the kerogen lenses distribution to the areas of the maximum pore accumulation of the given dimension in the sample 2 (Fig.4) that leads to the single direction of the stratification and pore accumulation areas.

Table 1. Statistic data on porous space of the samples.

No.	Parameter		Sample 1	Sample 2
1	The volumes of pores, mm^3	Min	5,22E-08	5,22E-08
2		Max	0,0230	3,02E-05
3		Mean	1,61E-07	1,02E-07
4		Sum	0,2553	0,2152
5	The volume of the sample, mm^3		21,0158	32,5160
6	Porosity, %		1,2147	0,6618
7	Pores number		1586798	2105755

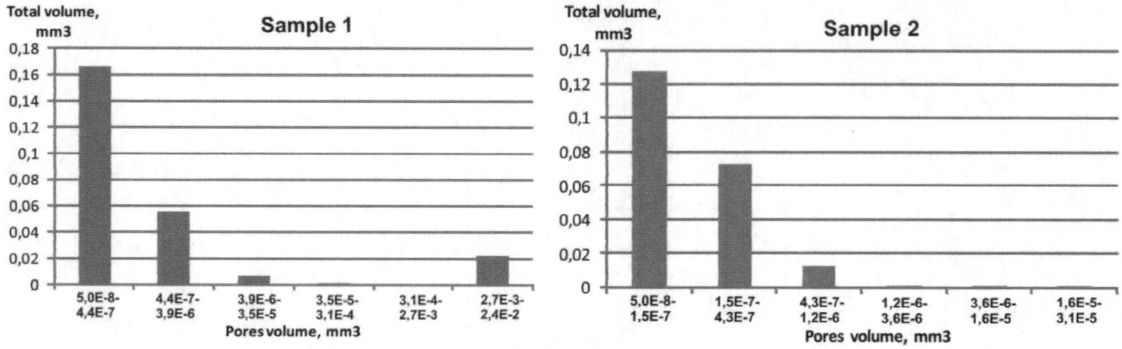

Figure 3. Pores distribution of different volumes in the rate of total volume for the samples.

For the sample 1, in the range 2.7×10^{-3}-2.4×10-2 (see Figure 3), a large volume is observed associated with the leaching crack (Fig. 5). In the authors'opinion, the oil shale intervals with such cracks are of the greatest interest to EOR, since they include a larger area with smaller pores, connecting them to each other. In addition, the pressurization, as well as the penetration and spread of various agents in the formation are occuring through them.

Figure 4. Correspondence of dark kerogen-saturated areas to the areas of small sizes pores in the sample 2. For this reason, the image shows an equal inclined direction of the thin stratification and pore distribution.

Figure 5. Volume visualization of the leaching crack in the volume of the sample 1.The range of voids between intra kerogen and fracture porosity is confined to pore-micro-cavernous and pore inter crystalline space [6]. As we know from the conducted studies, they are poorly represented in the studied samples.

4. Conclusions

As a result of the conducted studies, it is revealed that two types of porous space are common for the studied samples of the Domanik oil shales. The most common is the intra kerogen porosity with the pore volumes of 5×10^{-8} 1×10^{-6} mm^3. The volumetric reconstruction of the structure of this pore space shows that the voids are confined directly to micro lenses of organic matter. The second type of the found void is represented by leaching cracks. It is characteristic for more carbonate varieties of the Dominik oil shale with a spotted structure. The oil shale intervals with such cracks are of greatest interest to the EOR, since they consist of a large area with smaller pores and through which pressurization and spread of various agents are possible to occur in order to increase the oil recovery.

Acknowledgements

The work is performed according to the Russian Government Program of Competitive Growth of Kazan Federal University.

References

[1] Durand B. Kerogen: Insoluble Organic Matter from Sedimentary Rocks / B. Durand. – Editions Technip: Paris, 1980. – 519 p.
[2] Prishchepa O.V. Oil-and-gas bearing deposits of the Domanik type - a reserve to maintain hydrocarbon production in industrial regions / O.V. Prishchepa, O.U. Averyanova, A.M. Zharkov // Georesources. - 2013. - No.4 (54). - P. 18-22.
[3] Baykov N. M. Prospects for the Shale Oil Fields Development (Part 1) / N.M. Baykov, E.N. Baykova // Oil Industry. - 2013. - No.5 – P. 120-123.
[4] Tao Wan B.S. Evaluation of the EOR Potential in Shale Oil Reservoirs by Cyclic Gas Injection/ Thesis of master of science in petroleum engineering: Texas Tech University, 2013.
[5] Tom Alexander, J. B. 2011. Shale Gas Revolution. Schlumberger.
[6] Kalmykov G.A. The structure of the Bazhenov oil and gas bearing complex as the basis for the forecast of differentiated oil production. - Diss. For the competition uch. Art. d. g. Moscow, 2016, 391 p.

Geological criteria and geophysical methods of natural bitumen deposits preparation to the development

B V Uspensky[1], M Ya Borovsky[2], R F Vafin[1], S E Valeeva[1], R A Mudarisova[1]

[1] Institute of Geology and Petroleum Technologies, Kazan Federal University, Kazan, Russia
[2] LTD "Geofizservis", Kazan, Russia

Abstract. The article considers the provisions of the ontogenesis of the following factors in the formation of natural bitumen clusters in the Permian deposits of the Melekesskiy region: genetic, geodynamic, structural and hydrogeological.
It is shown that tectonically weakened zones and zones of Neogene incisions development are fixed by high-precision gravimetry in the form of intense local minima of gravity. A favorable factor contributing to the "strengthening" of anomalous geophysical effects is the coincidence of the locations of these geological section heterogeneities in the plan.
It is recommended at the stage of experimental-industrial operation a complex of geophysical methods for monitoring the processes of natural bitumen deposits development by means of secondary impact on the formation. High-precision magnetic, thermal and electrical prospecting in various modifications are used.

1. Introduction

Effective development of hydrocarbon deposits depends on taking into account the features of the geological structure, which are caused by the conditions of traps and the processes of minerals genesis.

There are essential factors of traps and natural bitumen formation. Clear boundaries between individual factors do not exist, i.e. their influence can affect both the formation of traps and deposits. The first is in different geodynamic manifestations of the oscillatory movement of the earth's crust, structureforming movements; palaeogeomorphology, sedimentation, litho-facies conditions, etc.; the second group includes factors of ontogenesis responsible for the generation, accumulation of hydrocarbons and the destruction of their clusters, i.e. geo(bio)chemical, hydrogeological, hypergenesis, etc.

2. Methodology

According to the ideas about the ontogenesis of oil are assumed [1-3], the following main factors of formation of natural bitumen clusters in the Permian deposits of the Melekesskiy region:

1. Genetic factor determines the generation of hydrocarbons. Formation of the Perm bitumen is associated with the presence of oil deposits in the Devonian and Carboniferous deposits (Fig. 1; 2). The main strata, with source potential in the Devonian system – clay-carbonate, domanic facies. In the above-lying lower Carboniferous rocks, the processes of oil and gas formation also took place in all negative structures.
Oil content of Permian deposits was provided by vertical migration of oil. The main source of hydrocarbons revenue are regional oil-bearing strata of Devonian and Carboniferous served, which is

IOP Conf. Series: Earth and Environmental Science **155** (2018) 012023 doi:10.1088/1755-1315/155/1/012023

confirmed by coincidence in terms of Permian bituminous areas with areas of concentration of oil deposits (Fig. 2)

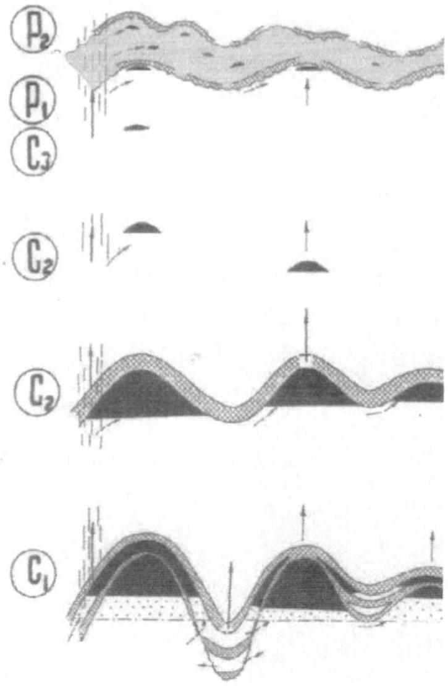

Figure 1. Scheme of oil deposits and natural bitumen accumulations formation in Carboniferous and Permian deposits (according to RS Khisamov, 2006)

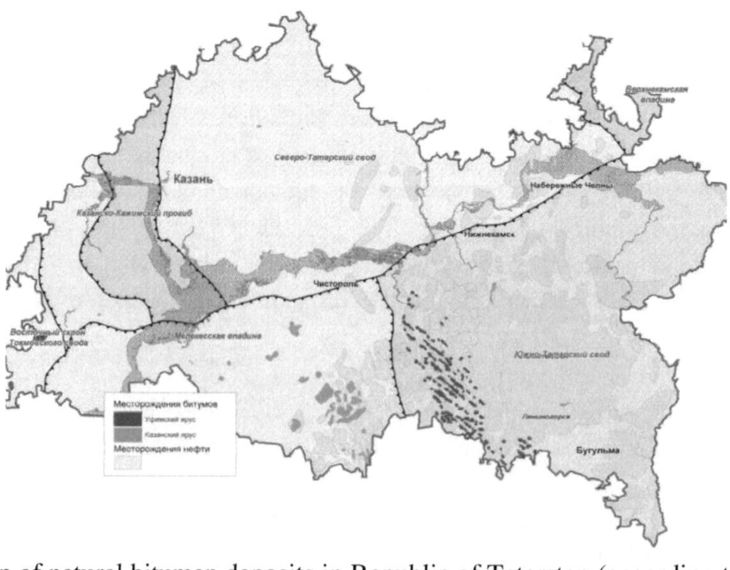

Figure 2. Map of natural bitumen deposits in Republic of Tatarstan (according to RKh, 1999)

2. Geodynamic factor characterizes modern and paleostructural plans, migration and accumulation of hydrocarbon deposits, phasing of formation.

Concentrations of natural bitumen of Permian deposits of Melekess depression formed by the lateral-vertical migration of hydrocarbons from more deeply buried Devonian and Carboniferous sediments. Ways of vertical flows are zones of tectonic disturbances, which provided, first of all, filling of the lower Permian reservoirs. The greatest development faults are at the junction of the South-Tatar arch, with its flanking valleys. On the Eastern side of the Melekess depression, this zone covers the entire complex of sedimentary formations. Migration of hydrocarbons in Sheshmian horizon went up the rise of the strata - eastward from the area of Melekess depression in the direction of the South-Tatar arch. Hydrocarbons enter Kazanian natural reservoirs as a result of gradual migration.

Modern and paleotectonic structural plans of the territory determine the conditions of hydrocarbon accumulation, the formation of their clusters.
Phasing of geotectonic development of the region. There were [3] several stages of generation, migration, accumulation of oil and destruction of its deposits, which determined the frequency of formation of oil deposits and its transformation into natural bitumen.

3. Hydrogeological factor manifests itself in two aspects:
1) in the formation of oil deposits - primary migration of hydrocarbon is carried out as part of the aqueous phase (aqueous molecular, mycelial, colloidal solutions, etc. [4]. The main reasons for the movement of oil in the reservoir rock are gravitational forces float lighter oil in the water, capillary forces and hydrodynamic water pressure. With the inclined position of the reservoir, the reservoir POPs up after the uprising of the reservoir until this pop-up is stopped (bending of the reservoir in the opposite direction, impenetrable screen, etc.). An important driving force in the process of secondary migration is the hydrodynamic pressure in the reservoir. This pressure helps the oil to float and helps to overcome the capillary forces in the narrowing of the pore channels;

2) when the deposits are destroyed. Under the influence of infiltration water, saturated with oxygen and microorganisms, processes hypergenesis. Plantar sulphate waters lead as a result of sulphate reduction to a change in the hydrocarbon composition of oil (bitumen) and, in general, to an increase in resinous-asphaltenic components. The formation of deep paleovalleys promotes the penetration of infiltration waters: the deeper erosion of the valley, and the greater the interval of the sedimentary cover affected by the processes of their impact. Long-term contact with oil deposits leads to oxidation of the latter and the formation of natural bitumen.

3. Results and Discussions

The results of the analysis of natural bitumen deposits formation conditions testify [1-3]:
1) Formation of Perm natural reservoirs occurred in different facies and paleogeographic conditions by successive changes in the conditions of sedimentation accumulation in the East of the Russian plate.
2) The initial stage of Permian sediments formation in some areas is associated with the covering of the nuclei of rhyme Carboniferous formations.
3) Many Perm structural forms are due to sedimentation processes mapped buried paleogeomorphological surfaces.
4) The structure of the pore space of the lower Permian oil sands reservoirs involved the pores diagenetic recrystallization and leaching, shallow cavities having predominantly isolated distribution; filtration capacity of the rocks determines the fracturing of carbonate rocks;

5) Traps of bitumen in Ufimian-Lower Kazanian are nontectonically. The coincidence of the structural plans of the roof of the sandstone pack and overlapping rocks is not associated with tectonic processes, and the consequence of uneven compaction processes lithologically heterogeneous rocks. The deposition of clastic material occurred in the mobile hydrodynamic setting with a cutout in the underlying rocks in deltaic and avandelta parts of paleorivers.
6) Deposits of sandstone pack form linear bodies, located in accordance with the flow of the river within the riverbed itself, delta and avandelta; allocated band (zone) elevations, separated from each other by depressions, within which the collector is characterized or sharply reduced (less than 10 m) capacity or completely absent.
7) The stretch of sandy bodies can be seen in the South-East direction outside Tatarstan in the Samara and Orenburg regions, where the transportation of terrigenous material (sandstone pack corresponds to the productive layers of the U-I and U -II in the roof of the Sheshmian horizon). In these areas the thickness of the formations increases, there are more coarse-grained differences of rocks. This suggests that the sandy sediments here are made of a paleorivers channel. Natural reservoirs of reservoir rocks are characterized by acies unevenness.
8) There is a certain sequence of formation and placement of lithographs.

On the basis of generalization of geological researches and pilot-industrial works results on extraction of heavy oils and natural bitumen it is revealed [1,2,5-8] a number of complicating factors which should be considered at the later stages of a geological exploration cycle:
1) Zones of increased tectonic fracturing of sedimentary thickness;
2) Valley of paleorivers (Neogene incisions);
3) Degree of destruction of bitumen deposit;
4) Hydrogeological features of the upper part of the cut.

4. Conclusions

Geophysical methods are used in the preparation of projects and technological schemes of industrial development of bitumen deposits at the exploration stage of mineral deposits. Lead is a highly mobile method of error-correcting gravity: traced the development zone and the Neogene incisions. It was found that the bitumen in the direction of the incision weighed down and in the zone are subjected to the most profound hypergenic effects. This phenomenon leads to a sharp deterioration in the commodity properties of hydrocarbons.

Gravity exploration acquires great importance as an effective method in the diagnosis of tectonic disintegration zones in the sedimentary thickness when solving a number of issues related to the processes of experimental and industrial exploitation of natural bitumen deposits by methods of secondary impact on the formation (intraplate moving center of combustion, steam injection, solvents, etc.). In case of extraction of hydrocarbons by in-situ technology in zones of increased tectonic fractures it can be observed premature breakthrough of heat carriers to the bottom of producing wells, or penetration of gaseous products of physic-chemical reactions into the covering deposit thickness up to the daily surface. Obviously, timely (at the stage of exploration) detection of zones of increased fracturing and permeability of layers can improve the efficiency of the development of hydrocarbon accumulations.

Tectonically weakened zones and zones of Neogene incisions development are fixed by high-precision gravimetry in the form of intense local minima of gravity. A favorable factor contributing to the "strengthening" of anomalous geophysical effects is the coincidence of the locations of these geological section heterogeneities in the plan.

At the stage of pilot operation, it is recommended to use a set of geophysical methods to control the processes of development of natural bitumen deposits by methods of secondary impact on the formation. It is applied [8,9] high-precision magnetic, thermal and electrical exploration in various modifications.

References

[1] Muslimov R Kh *et al* 2012 *Complex development of heavy oil and natural bitumen Permian system of the Republic of Tatarstan* (Kazan: FEN Publishing House of the Academy of Sciences of the Republic of Tatarstan) p 396

[2] Uspensky B V 2005 *Scientific and methodical bases of search, investigation and development of natural bitumen: the author. Diss. Doct. gay.- min. Sciences: 25.00.12* (Kazan: KSU) p 41

[3] Uspensky B V *et al* 2008 *Geology of the deposits of natural bitumen of Tatarstan* (Kazan: Publishing house PF GARTH) p 347

[4] Bazhenova O K *et al* 2000 *Geology and Geochemistry of oil and gas* (Moscow: Moscow state University Publishing house) p 384

[5] Borovsky M Ya 2001 *Geophysical technology of prediction, prospecting and exploration of natural bitumen deposits: Avtoref. Diss. kand. gay.- min. Sciences: 25.00.10* (Kazan: KSU) p 23

[6] Borovsky M Ya *et al* 2000 *Geophysical methods of preparation and control of processes of exploitation of natural bitumen deposits* (Moscow: GEOS) p 170

[7] Mukhametshin R Z 2006 *Geological basis for the effective development and extraction of hard-to-recover oil reserves: Avtoref. Diss. Doct. gay.- min. Sciences: 25.00.12, 25.00.17* (Moscow: Igirgi) p 51

[8] Khisamov R S *et al* 2007 *Geophysical methods of prospecting and exploration of deposits of natural bitumens in the Republic of Tatarstan* (Kazan: FEN Publishing House of the Academy of Sciences of the Republic of Tatarstan) p 247

[9] Shvydkin E K 1988 *Control of combustion front advance by means of ground magnetic survey in the development of shallow bitumen and viscous oil deposits: Diss. kand. technical Sciences: 04.00.12* (Moscow: Vniigeosystem) p 164

[10] Muslimov R Kh 1999 *Unconventional sources of hydrocarbons – further development of old oil producing regions* (Georesources **1**) pp 21 – 27

[11] Khisamov R S 2006 *Tectonic and petroleum geological zoning of the territory of Tatarstan* (Kazan: FEN Publishing House of the Academy of Sciences of the Republic of Tatarstan) p 328

IOP Conf. Series: Earth and Environmental Science **155** (2018) 012024 doi:10.1088/1755-1315/155/1/012024

Geochemistry of radioactive elements in bituminous sands and sandstones of Permian bitumen deposits of Tatarstan (east of the Russian plate)

A I Mullakaev[1], R R Khasanov[1], O R Badrutdinov[2], I R Kamaletdinov[1]
[1] Institute of Geology and Petroleum Technologies, Kazan Federal University, Kazan, Russia
[2] Institute of Environmental Sciences, Kazan Federal University, Kazan, Russia

E-mail: rinat.khassanov@kpfu.ru, almazmullakaev@gmail.com

Abstract.

The article investigates geochemical features of Permian (Cisuralian, Ufimian Stage and Biarmian, Kazanian Stage of the General Stratigraphic Scale of Russia) bituminous sands and sandstones located on the territory of the Volga-Ural oil and gas province (Republic of Tatarstan). Natural bitumens are extracted using thermal methods as deposits of high-viscosity oils. In the samples studied, the specific activity of natural radionuclides from the ^{238}U (^{226}Ra), ^{232}Th, and ^{40}K series was measured using gamma spectrometry. As a result of the precipitation of uranium and thorium and their subsequent decay, the accumulation of radium (^{226}Ra and ^{228}Ra) has been shown to occur in the bituminous substance. In the process of exploitation of bitumen-bearing rock deposits (as an oil fields) radium in the composition of a water-oil mixture can be extracted to the surface or deposited on sulfate barriers, while being concentrated on the walls of pipes and other equipment. This process requires increased attention to monitoring and inspection the environmental safety of the exploitation procedure.

Key words: bitumen sands, bitumen sandstones, tar sands, radioactive elements, concentration

1. Introduction.

Within the Volga-Ural oil and gas province, there are deposits of natural bitumen [1,2,3], dated to the Ufimian and Kazanian deposits of the Permian period. Reduction of the reserves of light oil facilitates the introduction of new technologies for extracting hard-to-recover hydrocarbons, including natural bitumen. In the territory of the Republic of Tatarstan, the resources of bitumen-bearing rocks vary from 1.5 to 10 billion tons. More than 150 deposits and manifestations of natural bitumens are confined to Ufimian and Kazanian deposits, where they occur from the Earth's surface to depths of 400 m. At present, efforts are being made in

IOP Conf. Series: Earth and Environmental Science **155** (2018) 012024 doi:10.1088/1755-1315/155/1/012024

Tatarstan to develop bitumen deposits as high-viscosity oils. The complexity of their development is due to the fact that in the natural state they do not have the necessary parameters of viscosity and density to be extracted by traditional well methods. Thermal methods are used to achieve the necessary production conditions. Bituminous rocks are warmed by heated steam [3]. As a result, the bitumen substance acquires fluid properties and is extractable to the surface. In this connection, the study of the nature of the chemical changes occurring in the formation during the interaction of steam and oil fractions, as well as the trace element composition of the bitumen-bearing rocks, is of particular importance. The change in physical and chemical conditions of the environment for recovered hydrocarbon fractions during the exploitation of oil fields causes an undesirable concentration of metal compounds in the pipes.

One of the important geochemical problems in petroleum geology is an uncontrolled concentration of radioactive elements and, mainly, radium [4]. The nature of radium concentration is as follows. The most common isotopes in natural environments (sediments, rocks) are uranium 238 and thorium 232. Radium 226 is a product of the decay of uranium 238, which is mobile in an alkaline environment and can concentrate on acid barriers. In oil-bearing deposits uranium is concentrated in the zone of water-oil contact, where as a result of its decay, the concentration of radium 226 occurs. Part of the radium is also accumulated as a result of sedimentation from the groundwater of the enclosing sediments. During the exploitation of the oil field, radium in the water-oil mixture enters the surface, where it is deposited on sulfate barriers in the composition of sulfate compounds. As a result, areas with increased radioactivity may be formed, causing environmental problems.

The purpose of this work is to study the activity of natural radionuclides in bitumen-bearing rocks.

2. Methodology

To solve environmental problems arising during the exploitation of bitumen-bearing rocks by thermal methods, was made mineralogical and geochemical research of bitumen-containing rocks of the Sheshmian horizon (Ufimian Stage of Permian Period). Petrographic research has been carried out using methods of traditional polarization microcoscopy in transmitted light. To determine the composition of the finely dispersed mineral substance of cement and the type of organic matter, the electron paramagnetic resonance (EPR) method has been used. EPR [5,6] studies diamagnetic crystals in which the paramagnetic centers are impurity ions of Fe^{3+}, Mn^{2+} in calcite, dolomite, electron-hole centers in them, free radicals of organic matter of coal and oil series. These paramagnetic centers reflect the conditions and environment of formation of minerals, the degree of degradation of the syngenetic material of rocks, reflecting different stages of catagenetic transformation of rocks. In calcite ions Mn^{2+} are replacing calcium ions (Ca^{2+}), and in the dolomite they replace ions Ca^{2+} and Mg^{2+} in two structurally nonequivalent positions. In bitumen-containing sands and sandstones, the content

IOP Conf. Series: Earth and Environmental Science **155** (2018) 012024 doi:10.1088/1755-1315/155/1/012024

of natural radionuclides from the ^{238}U (^{226}Ra) series, as well as ^{232}Th and ^{40}K, was estimated by gamma spectrometry. The specific activity of natural radionuclides is measured in Bq / kg. Analyses were performed in the Institute of Environmental Sciences, KFU. The content of uranium and thorium was determined using ICP-MS in the laboratory of the Institute of Geology and Petroleum Technologies. The results of the analyses are shown in table 1.

Table 1

Results of gamma spectrometry and ICP-MS

№	Sample	Lithology types	^{40}K, Bq/kg.	^{226}Ra, Bq/kg.	^{232}Th, Bq/kg	U, ppm	Th, ppm
1	C 5	bituminous cemented sandstones	190,00	9,68	N/A	N/A	0
2	C 8	bituminous cemented sandstones	235,60	8,93	N/A	N/A	1,36
3	C 13	bituminous cemented sandstones	170,70	N/A	9,40	N/A	N/A
4	C 16	bituminous cemented sandstones	199,40	9,36	N/A	N/A	1,48
5	C 21	cemented sandstones	168,00	3,24	N/A	N/A	N/A
6	C 22	cemented sandstones	179,10	6,28	N/A	N/A	1,47
7	C 28	cemented sandstones	167,70	5,44	N/A	N/A	N/A
8	C 35	cemented sandstones	358,00	6,59	7,98	N/A	2,29
9	D 1	cemented sandstones	193,90	7,27	6,41	N/A	N/A
10	D 3	bituminous cemented sandstones	166,60	6,18	N/A	N/A	N/A
11	D 7	tar sands	212,10	13,30	N/A	N/A	1,63
12	D 10	tar sands	193,50	29,53	N/A	N/A	N/A
13	D 11	bituminous cemented sandstones	196,20	6,05	N/A	N/A	N/A
14	D 14	bituminous cemented sandstones	206,20	5,96	N/A	N/A	1,36
15	D 19	tar sands	206,30	5,21	N/A	2,89	3,92
16	D 24	bituminous cemented sandstones	187,00	12,18	N/A	N/A	1,66
17	D 27	bituminous cemented sandstones	266,00	7,12	N/A	1,49	1,6
18	D 31	cemented sandstones	268,70	15,47	N/A	N/A	N/A
19	D 32	cemented sandstones	167,20	7,54	N/A	2,94	9,35
20	D 38	cemented sandstones	239,80	6,43	4,73	N/A	1,98
21	D 46	cemented sandstones	193,70	6,02	N/A	18,18	N/A

N/A-data not available or below the detection limit of the device

3. Results and Discussions

According to the mineral composition, the investigated sandstones are polymictic and belong to the greywack group [1,7]. In the clastic material of sandstones, there can be found grains of quartz, feldspar, mica, as well as particles of magmatic rocks, the source of which were the magmatic rocks of the Ural Mountains destroyed in the Permian period. Particles of acidic rocks predominate among the magmatic fragments. The clastic material is not sufficiently sorted out, it is dominated by uncoated, angular fragments that indicate the short duration of the process of their washing and transfer [8]. The rocks of the Sheshmian horizon are represented by loose uncemented varieties (sands) and dense, hard-cemented sandstones. The cement in dense sandstones is clayey-carbonate, sometimes carbonate, proportion of which in sand and sandstones can be more than 30%. The presence of clayey components and carbonate minerals that were difficult to determine in the sandstones was confirmed by the EPR method. The EPR spectra in the rock revealed Fe3+ signals associated with clay minerals, as well as Mn2+ signals typical of dolomite and calcite. An important component of the studied sandstones is the organic matter, the forms of its presence in rocks are also confidently fixed by the EPR method [9]. There are two types of organic matter - the oil and coal series in the studied samples. The oil substance has a migration nature and, apparently, is associated with an influx of water-oil fluid from deeper horizons (Carboniferous sediments). The organic matter of the coal series is the syngenetic remains of plants deposited together with the primary sediments. In loose sands the role of cement is played by a bitumen substance that envelops fragments of quartz, feldspar and other minerals. The element composition of the organic matter in the samples studied has the values: C (72.05-83.31%), H (10.08-11.62%), N (0.20-0.50%) [7]. The bulk of the samples is characterized by similar parameters which indicates the homogeneity of the composition of bitumen.

Due to the scale of their presence, the ^{238}U (^{226}Ra) and ^{232}Th (^{228}Ra) and ^{40}K radionuclides have a significant influence on the geochemical behavior of the petroleum substance, which is unevenly distributed in the studied rocks [11]. Depending on the content of organic matter, the rocks studied were divided into three groups - tar sands, bituminous cemented sandstones and cemented sandstones [8].

The structure of bituminous deposits is characterized by vertical zoning. The horizons of loose sands are located mainly in the upper sections of the productive horizon, and strong differences are in the lower horizons. This pattern can be explained by cementation of rocks below the zone of water-oil contact [12], where geochemical contrasts arose between the oil waters and the waters of the enclosing strata. Secondary mineral formation in the zone of water-oil contact was accompanied by cementation of the parent rocks with calcite (and dolomite). The underground water of the underlying sulfate-carbonate Early Permian deposits are quite likely to be the source of calcium for the crystallization of dispersed calcite in the porous space of sandstones. The groundwater is characterized by favorable for the migration of calcium and magnesium alkaline properties.

Different types of studied sandstones are characterized by different activity of the radionuclides considered, which is explained by their different geochemical behavior and forms of location. It is known [13] that uranium in sedimentary rocks is mainly associated with organic matter and partly with apatite and calcite, thorium with clay minerals, and potassium with clay minerals and potassium feldspar. The correlation of radioactive elements between each other indirectly reflects their connections with the mineral components of rocks (Table 2). However, it should be noted that due to the low content of uranium and thorium in the rocks, the correlation bonds are weakly expressed and can be used only to reveal general patterns.

IOP Conf. Series: Earth and Environmental Science **155** (2018) 012024 doi:10.1088/1755-1315/155/1/012024

Table 2

Correlation matrix of specific activity of radionuclides and concentrations of uranium and thorium

	^{40}K	^{226}Ra	^{232}Th	U	Th
^{40}K	1,00	0,10	0,33	-0,08	0,03
^{226}Ra	0,10	1,00	-0,32	-0,12	-0,09
^{232}Th	0,33	-0,32	1,00	-0,15	-0,08
U	-0,08	-0,12	-0,15	1,00	0,04
Th	0,03	-0,09	-0,08	0,04	1,00

In the rocks being studied, we can identify association of potassium with sands and sandstones greatest values of which occur in bitumen-bearing varieties and decrease in clays and clayey rocks. The reason for this is that potassium is found primarily in potassium feldspars, and potassium feldspars are a constant component of the clastic material of the examined sandstones. An increase in the values of the specific activity of potassium in rocks with organic matter should be attributed, apparently, to its precipitation on acid barriers. And the origin of the reactive dissolved potassium - with the processes of leaching in the zones of water-oil contact.

Thorium (^{232}Th) is a natural radionuclide. Thorium does not form elevated concentrations in the rocks under consideration (table 1). Some increase in its concentrations is associated with the sites of rock carbonatization. In nature, it occurs mainly in acidic magmatic rocks, while in sedimentary rocks it concentrate in combustible shales [13]. It can be assumed that the introduction of the bulk of thorium into the Ufimian deposits occurred as part of fragments of igneous rocks, and it enters the composition of groundwater during the decomposition of carrier minerals. In the rocks under consideration, its enhanced correlation with Uranium is noted. Apparently, this is due to its deposition in small quantities on acid barriers in the zone of water-oil contact, where it accumulates along with uranium on cementation zones.

A subject of great interest is radium characterized by a non-uniform distribution with increasing specific activity in rocks containing hydrocarbon fractions, especially in tar sands with bitumen (Table 1). In cemented rocks, its activity is much lower. Radium is a product of the decay of uranium and thorium. The uranium content in sands and sandstones is low and is within the clarke of concentration for sedimentary rocks (2-4 ppm) and below the sensitivity threshold of the method. Its concentration is connected, apparently, with acidic barriers in the zone of water-oil contact. Radium accumulates in oils as a result of the constant addition of uranium from surrounding rocks and its further decay. The negative bonds of radium (^{226}Ra) with uranium concentrations should be attributed to a decrease in uranium concentrations as a result of its decay and the increase in activity of the resulting radium.

4. Conclusions

Summarizing all the above, we can state the following. In the zone of water-oil contact, where potassium is affected by groundwater, potassium is likely to be dissolved and be carried over with redepositing in the other parts of the section. Such areas are could be zones of sandstone cementation where mass calcite precipitation has been found to occur. Links with other radionuclides, especially with thorium, are not explicit. This is due to the fact that the content of radionuclides in different parts of the section is affected by oil and associated waters.

It appears to occur the accumulation of radium (^{226}Ra and ^{228}Ra) as a result of the decay of uranium and thorium in bitumen substance. Thus, in the process of exploitation of deposits of bitumen-containing rocks as deposits of highly viscous oils, phenomena similar to those observed in oil fields may occur [4]. Radium constituting the water-oil mixture can be extracted to the surface and deposited on sulfate barriers concentrating on the walls of pipes and other equipment. This process requires increased attention to monitoring and inspection the environmental safety of the exploitation procedure.

This study was funded by the subsidy allocated to Kazan Federal University as part of the state program for increasing its competitiveness among the world's leading centers of science and education.

References

[1] Troyepolskiy V.I., Lebedev N.P. Produktivnyye bituminoznyye tolshchi permskikh otlozheniy Melekesskoy vpadiny i Tatarskogo svoda. Kazan': Izdatel'stvo Kazanskogo universiteta, 1982. – 104.
[2] Uspenskiy B.V., Valeyeva I.F. Geologiya mestorozhdeniy prirodnykh bitumov Respubliki Tatarstan. Kazan': Izdatel'stvo OOO "PF "Gart", 2008. - 348.
[3] Geologiya i osvoyeniye zalezhey prirodnykh bitumov Respubliki Tatarstan / R.S. Khisamov, N.S. Gatiyatullin, I.Ye. Shargorodskiy i dr. - Kazan': izd-vo "Fen", AN RT, 2007. – 295.
[4] Ibragimov N.G., Sizov B.A., Zajtsev, Khisamutdinov A.F., Kubarev P.N. Radiation safety of the personnel and environmental contamination prevention at oil recovery and treating//Neftyanoe Khozyaystvo - Oil Industry. - 2005. - Vol., Is.3. - P.12-14.
[5] *Bulka G.R., Nizamutdinov N.M., Mukhutdinova N.G. et al.* EPR Probes in Sedimentary Rocks: The Features of Mn2+ and Free Radicals Distribution in the Permian Formation in Tartarstan // Appl. Magn. Reson. – 1991. – V. 2. – P. 107–115.
[6] *Ikeya M.* New Applications of Electron Spin Resonance- Dating, Dosimetry and Microscopy // World Scientific. – Singapore, 1993. – 509 p.

[7] Khasanov R.R., Mullakaev A.I., Dusmanov E.N. The structure of sandstones in productive horizons of the Permian bituminous deposits of Tatarstan (Russia). Uchenye Zapiski Kazanskogo Universiteta. Seriya Estestvennye Nauki, 2017, vol. 159, no. 1, pp. 164–173. (In Russian)
[8] Mullakaev, A.I, Khasanov, R.R., Galiullin, B.M. Mineralogy of sandstones and localization of oil matter in productive horizons of high-viscosity oil in permian deposits of the Volga-Ural region (Russia)/ 17th International Multidisciplinary Scientific GeoConference SGEM 2017, SGEM2017 Conference Proceedings, 29 June - 5 July 2017, Vol. 17, Issue 11, 353-358 pp. DOI:10.5593/sgem2017/11.
[9] *Conard J.* EPR in fossil carbonaceous materials // Magnetic Resonance. Introduction, Advanced Topics and Application to Fossil Energy / Eds. L. Petrakis, J.P. Fraissard. – Dordrecht: Reidel, 1984. – P. 441–459.
[10] *Jezierski A., Czechowski F., Jerzykiewicz M., Drozd J.* EPR Investigations of Structure of Humic Acids from Compost, Soil, Peat and Soft Brown Coal upon Oxidation and Metal Uptake // Appl. Magn. Reson. – 2000. – V. 1, No 18. – P. 127–136.
[11] Mullakaev, A.I, Khasanov, R.R., Badrutdinov O.R., Khasanova N.M., Nizamutdnov N.M. Features of bitumen-containing sandstones of the Volga-Ural oil and gas province according to Electron paramagnetic resonance and Gamma spectrometry data / 17th International Multidisciplinary Scientific GeoConference SGEM 2017, SGEM2017 Conference Proceedings, 29 June - 5 July 2017, Vol. 17, Issue 11, 133-139 pp. DOI:10.5593/sgem2017/11 DOI: 10.5593/sgem2017/11/S01.017
[12] Sakhibgareev R.S. Secondary Changes in Reservoirs during the Formation and Destruction of Oil Deposits. Leningrad, Nedra, 1989. 260 p. (In Russian)
[13] Y.N Zanin, A.G. Zamirailova, V.G. Eder. Uranium, thorium, and potassium in black shales of the Bazhenov Formation of the West Siberian marine basin Lithology and Mineral Resources. 2016. T. 51. № 1. C. 74-85. DOI: 10.7868/S0024497X16010079

IOP Conf. Series: Earth and Environmental Science **155** (2018) 012025 doi:10.1088/1755-1315/155/1/012025

Mobile geophysics for searching and exploration of Domanic hydrocarbon deposits

M Ya Borovsky[1], B V Uspensky[2], S E Valeeva[2], A S Borisov[2]

[1]LTD "Geofizservis", Kazan, Russia
[2] Institute of Geology and Petroleum Technologies, Kazan Federal University, Kazan, Russia

Abstract. There are noted features of shale hydrocarbons occurrence. It is shown the role of geophysical prospecting in the geological prospecting process for non-traditional sources of hydrocarbon. There are considered the possibilities of non-seismic methods for forecasting, prospecting, exploration and preparation of Domanikovian hydrocarbons accumulations for exploration. It is emphasized the need for geophysical studies of tectonic disturbances. Modern aerogeophysical instrumentation and methodological support allows to combine high-precision magneto-prospecting with gravimetric and gamma spectrometry. This combination of geophysical methods contributes to the diagnosis of active and latent faults.

1. Introduction

In the Volga-Urals oil and gas basin, Zaidelson et al [1] distinguish "domanikites", representing bituminous clayey-siliceous carbonates of Upper Devonian-Tournaisian deposits. In the Russian Federation, these rocks, along with the Bazhenov Formation and the Hadum Formation of the Ciscaucasia, are attributed to the shale sequences [2].

There are the features of the hydrocarbons accumulations in shale [3, 4]:

- shales - the source rocks and the reservoir at the same time;
- low porosity and permeability of the enclosing strata;
- diffuse state of hydrocarbons in the stratum;
- continuity of clusters;
- lack of traditional systems for monitoring deposits (oil-water contact, structural, lithological).

2. Methodology

A decisive role at different stages of geological exploration for non-traditional sources of hydrocarbons, including shale oil and gas, belongs to geophysical exploration. The extraction process requires modern technologies for both prospecting and exploration of deposits, and for the implementation of the developing the desired objects process. This necessitates a high-level real-time visualization of processes in the shale reservoir, including changes in the physical and mechanical characteristics and parameters of the reservoir rock, the state of the deposit during production. A geophysical observation, in which seismic prospecting plays an important role [5] serves as a tool for technical vision in solving the problem of prospecting, exploration, development and production of shale hydrocarbons.

The efficiency of the geological exploration process depends to the choice of research directions resulting from the study of each particular facility. Separate stages and stages of exploration work have as their goal the study of specific large-scale objects.

At the initial stage of sedimentary basins or their separate parts research, regional work is conducted to study the inherent common features of the geological structure and oil and gas. In the future, at different stages of the basins study, regional problems are posed with specific problems, determined by the degree of prior study, by the technical and methodological capabilities of geological and geophysical studies.

Addressing issues at the regional level, as a tectonic zoning, is of the utmost importance for the search for minerals. It allows [6]:
1) to establish general prospects for the search minerals in the region;
2) identify areas that require more detailed search operations.

Khain stresses "... knowledge of the regional structure of an area is no less (if not more) important for assessing the prospects of oil than the significance of the local structure of its individual sections."

This provision is relevant for the general system of geological exploration for shale hydrocarbons.

Expressness and the current availability of a large number of maps and geophysical parameters graphs necessitate the use of aerogeophysical reconnaissance in information support of geological exploration for shale hydrocarbons. Essential as a choice of the most promising directions of "Exploration play" [7] and certain preparation of areas for detailed works.

The regional direction assumes [8, 9] the most complete use of geophysical surveys performed earlier in significant areas, which is due to the large area of shale gas and oil plays.

It is necessary to obtain information about the internal structure of the folded base to identify areas with specific sedimentation conditions against the background of surrounding areas,

3. Results and Discussions

By the present time, certain experience and ideas on the geophysical study of the crystalline basement have been accumulated [1, 2, 3]. Most researchers note that the structure and composition of the basement rocks is the main geological factor determining the nature of gravitational and, in particular, magnetic anomalies. An obligatory element of the method is the compilation of physical fields zoning maps or schemes.

Analysis of geological and geophysical materials indicates the possibility of isolating regional tectonic disturbances by the geophysical complex (gravimetry and magneto-prospecting) and remote (space survey) methods with data on the landscape indication. These regional tectonic disturbances are the boundaries on which the facies and thicknesses of various intervals of the sedimentary cover deposits.

For the most part the criteria for detecting and tracing faults from the results of geophysical observations are generally accepted [1-5]: the boundaries of sections with specific morphology of the geomagnetic field and the gravity field; zones of high horizontal gradients (magnetic and gravitational steps); steep and sharp limitation of magnetic anomalies. The peculiarity of the regional disruptive disturbances manifestation is the presence of secondary deformations of the gravitational (in zones of intense horizontal gradients) and magnetic fields. The study of the geophysical fields structure makes it possible in many cases to determine the character of the tectonic movements along the fault zones.

When preparing prospective areas for detailed geological prospecting for hydrocarbons of shale, attention should be paid to the compilation of a structural skeleton of fault tectonics. Analysis of faults, in particular their stresses and orientations, is necessary [5] for both regional and detailed scale of reservoir interpretation: faults and discontinuities can dynamically affect. It should be noted [7] that clarifying the position of the faults and assessing their effect on the conditions for the formation of hydrocarbon traps is one of the main tasks of aerogeophysical exploration.

It is advisable to use high-precision gravimetric prospecting at the stages of local forecasting, which makes it possible to determine the distribution in the cross-section and lateral, such a parameter as density, and, consequently, porosity. New principles of interpretation based on gravity modeling [10] and tomographic processing and visualization [11] give (Fig. 1) extensive information on decompressed and compacted (low-porous, low-permeability) zones and areas.

By Slepak [10] "... In studying the geological structure of oil fields, gravity prospecting should be aimed at identifying and interpreting the anomalies created by the lateral variability of densities in the sedimentary cover and in the crystalline basement." As a result of these studies, is developed a technique for solving inverse problems aimed at detecting anomalous changes in the gravitational field caused by lateral variability in densities. The fundamental point of the developed methods is the consideration of the earth crust as a two-layer model, consisting of a sedimentary cover and a crystalline base. Interpretation of the first upper layer makes it possible to study the features of the geological structure of oil deposits at sites of local structural forms and carry out their prediction, the second layer - to identify the block structure of the crystalline basement, the zone of faults and decompression of rocks.

To obtain reliable materials, according to Mikhailov [1-5] the following requirements must be met:
- straightness and parallelism of profiles;
- orthogonal network of profiles;
- the scale of the survey corresponds to the size of the search objects;
- the output of the gravimetric profile into the normal field of gravity (that is, beyond the limits of the anomalies-forming object).

Carrying out high-precision gravimetric works in combination with 3D seismic acquisition allows creating seismic-gravimetric models [10], which significantly improves seismic performance in forecasting and prospecting for oil deposits.

IOP Conf. Series: Earth and Environmental Science **155** (2018) 012025 doi:10.1088/1755-1315/155/1/012025

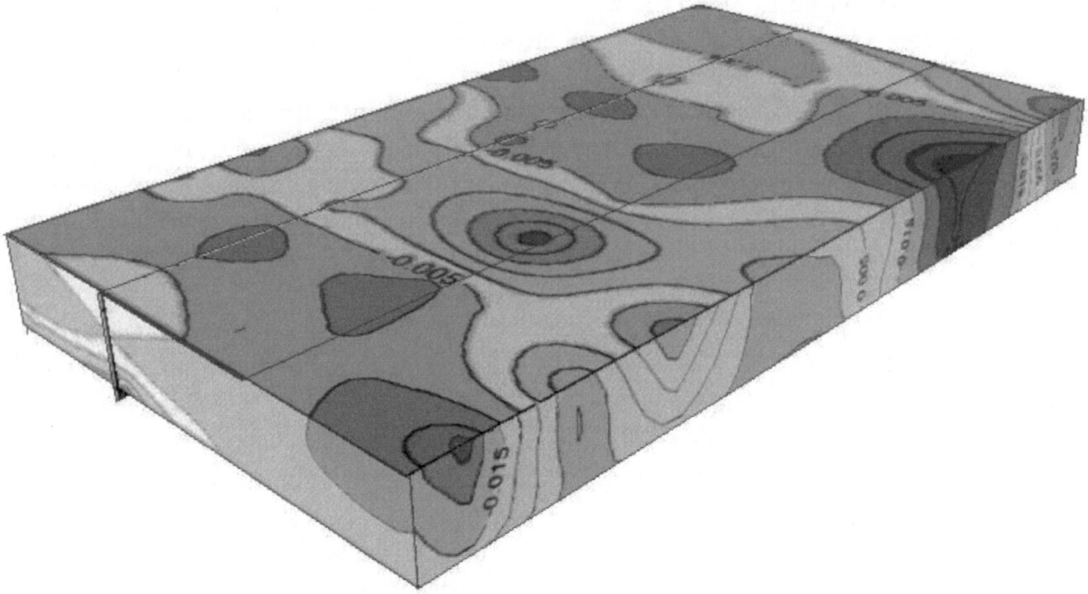

Figure 1. A volumetric model of the lateral variability of the densities in the section Yamashinskaya structure (according to Slepak, 2012)

The use of high-precision gravimetric prospecting will likely contribute to increasing the efficiency of prospecting for plots that are promising for shale hydrocarbons, at the stage of exploration and preparation for development – for optimal placement of horizontal wells.

There are only single descriptions of the results of the electrical exploration application for prospecting and exploration of shale hydrocarbons deposits. It is conducted [12] experiments on the feasibility of application non-traditional geoelectric methods for the formation of a short-pulse electromagnetic field (SPEF) and vertical electroresonance sounding (VERS) and a new method for processing remote sensing data (RS) to search for free gas accumulations (methane) within the coal-bearing rocks. By an area survey using the SPEF method, anomalies such as a "free gas (methane) deposit" or an oil deposit can be detected and mapped (Fig. 2). Depths of the anomalously polarized formations (APF) location of the "gas" type are determined by the VERS probing. The method of processing remote sensing data also allows the rapid identification and mapping of anomalies such as "free gas deposits". The results of the experiments, in the opinion of the authors [12] show that the SPEF-VERZ technology, together with the remote sensing data processing method, can be successfully used in the search and exploration of free gas accumulations (methane) within the distribution of coal-bearing formations.

Shell's experience shows [12] that only 5% of shale spreading areas are suitable for economically acceptable gas production. The use of the proposed express technology of "direct" searches for hydrocarbon accumulations shows [12] that the identified anomalies of "hydrocarbon reservoir type" occupy about 6% of the area searches. As a result, it is concluded that mobile technologies allow the prompt discovery and mapping of the "Sweet Spots" zones within which industrial development of shale hydrocarbons can be organized.

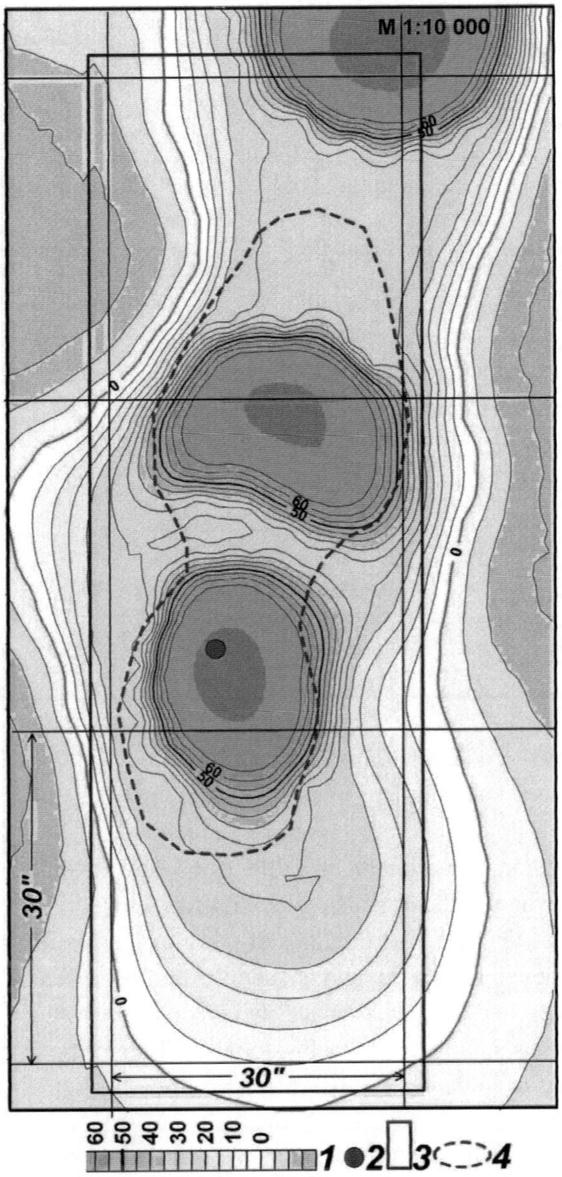

Figure 2. The "hydrocarbon reservoir type" map of the "oil" type
1 - scale of intensity of response in pressure values (MPa); 2 - drilled well, 3 - contour of the site; 4 - contour of "hydrocarbon reservoir type" type "oil" (according to SP Levashov et al., 2013)

There are proposals to use the biogeophysical method (BGF-method) for the search and exploration of shale hydrocarbons, the bases of which were developed by the Russian geophysicist N.M. Andreev. The technology for prospecting and exploration of hydrocarbon deposits based on the BGF method was described by the author at the All-Russian Conference on Deep Genesis of Oil "1st Kudryavtsev Readings" (Moscow, CGE, October 22-25, 2012).

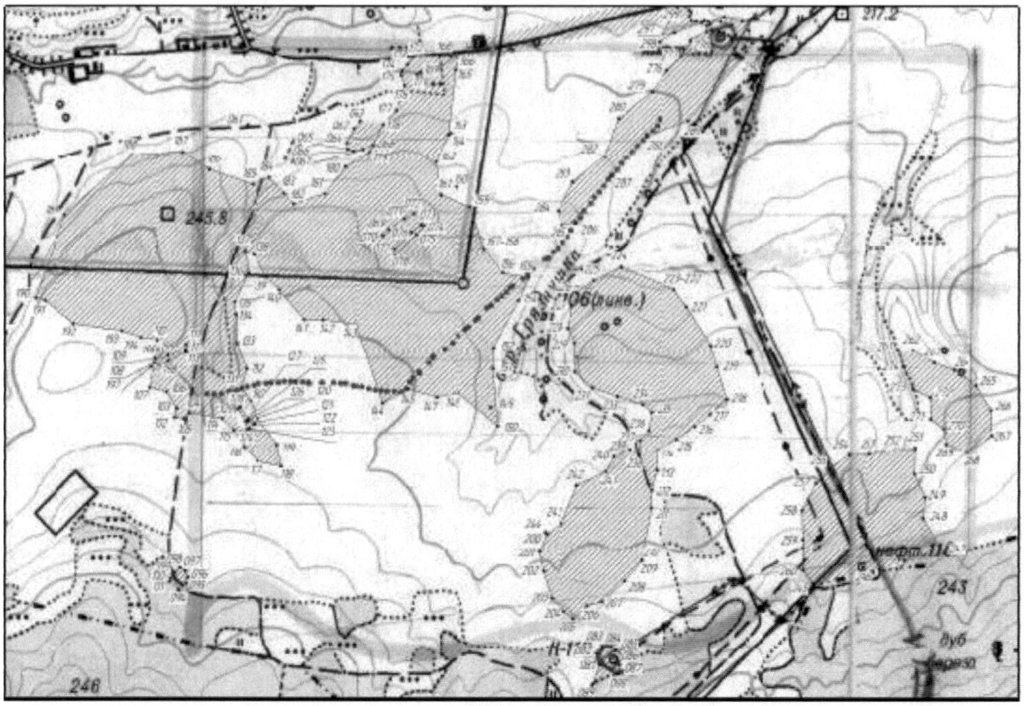

Figure 3. The results of a detailed mapping of extensive PGF anomalies - oil deposits (according to NM Andreev, 2012)

At the base of the phenomenon is the interaction effect of a certain field of the Earth with a highly sensitive biological sensor, which is the organism of the human operator. The effect of this field on such a sensor, apparently, changes the vector of the lines of force of the so-called "human biofield", which leads to rotation of the frames in his hands, giving them certain positions. At the same time, the force acting on the frame and the position it takes have its own characteristic features when the operator crosses various manifestations of the given field-global grid structures that are his background or boundaries of anomalies that in some places disturb the harmonious picture of the grid structures. This allows (according to NM Andreev) to isolate and map (Fig. 3) such anomalies on the surface of the Earth, which is the main task of the BGF method, in contrast to dowsing, radio-esesia, etc.), whose adherents claim the possibility "Dialogue" with a certain "information field" of the Earth with the help of frames (vines), receiving very different information about the objects of the search.

The use of direct searches for hydrocarbons in the enclosing strata is complicated [1, 3], apparently, by the scattered nature of the hydrocarbons occurrence in shales. Probably, concentrated accumulations ("deposits") of shale oil and gas are formed as a result of a fracturing operation, when hydrocarbons are drained along the cracks to a horizontal well. In other words, the formation of separate deposits of shale hydrocarbons is technogenic in our opinion.

The process of crack formation after hydraulic fracturing of the productive formation is fixed [5] with microseismic observations. There are [13, 14, 15] very favorable physico-geological prerequisites for the use of high-resolution electrical prospecting. The development of shale oil and gas necessitates the prediction of [1-5] migration routes of concentrated fluid streams formed as a result of a fracturing operation of a productive formation in a horizontal well and entering the upper layers of the sedimentary cover. The presence of weakened zones affects the efficiency of hydraulic fracturing and creates the danger of negative impact on the environment.

4. Conclusions

Modern aerogeophysical instrumentation and methodological support allows [7] to integrate high-precision magneto-prospecting with gravimetric and gamma spectrometry. This combination of geophysical methods contributes to the diagnosis of faults to active and latent ("dead").

The development of new technologies for locating zones of high fracturing and continuity, identifying zones of oil, gas and water saturation, tracking the dynamics of fluid flows cause [16] the use of monitoring methods that provide information about the changed properties , geometry and structure of productive horizons.

Acknowledges
This work was supported by the Russian Foundation for Basic Research (Project No. 18-55-53051).

References

[1] Zaidelson M I *et al* 1990 *Formation and oil and gas content of domanicoid formations* (Moscow: Nauka) p 79
[2] Arutyunov T V *et al* 2015 *Analysis of the characteristics of the geological structure and reservoir properties of shale deposits* (*Oil. Gas. Innovations* **4**) pp 72-76
[3] Prischepa O M *et al* 2013 Oil shale of the East European platform - a reserve for maintaining hydrocarbon production in industrialized areas *Problems of increasing the efficiency of development of oil deposits at a late stage: Proceedings of the International Scientific and Practical Conference* (Kazan: FEN Publishing House of the Academy of Sciences of the Republic of Tatarstan) pp 28-32
[4] Grigoriev G A *et al* 2012 *Prospects of industrial development of non-traditional gas resources in Russia* (*Oil and gas geology. Theory and practice*. vol 7 **2**) pp 1-19
[5] Goshevsky *et al* 2012 *Development of new geophysical technologies for the exploration and development of shale gas* (Collection of scientific works of UkrDGR **1**) pp 9-32
[6] Borovsky M Ya 2001*Geophysical technology of forecasting, prospecting and exploration of deposits of natural bitumens*: the author's abstract. diss. (Kazan: KSU) p 23
[7] Borovsky M Ya *et al 2014 Aerogeophysical exploration for hydrocarbons of shale formations* (*Oil. Gas. Innovations* **6**) pp 16-19
[8] Khisamov R S *et al 2007 Geophysical methods of prospecting and exploration of natural bitumen deposits in the Republic of Tatarstan* (Kazan: FEN Publishing House of the Academy of Sciences of the Republic of Tatarstan) p 247
[9] Khisamov R S *et al 2010 Geological exploration in regions with high prospecting of subsoil* (Kazan: FEN Publishing House of the Academy of Sciences of the Republic of Tatarstan) p 274
[10] Slepak Z M 2012 Gravity prospecting for forecasting natural bitumen and high-viscosity oils *High-viscosity oil and natural bitumen: problems and increasing the efficiency of exploration and development of deposits: Proceedings of the International Scientific and Practical Conference* (Kazan: FEN Publishing House of the Academy of Sciences of the Republic of Tatarstan) pp 106-109

[11] Bychkov S G 2010 *Methods of processing and interpretation of gravimetric observations in the solution of problems of oil and gas geology* (Ekaterinburg: UrB RAS) p 187
[12] Levashov S P *et al 2012* Detection and mapping of gas accumulations in shale rocks by mobile geophysical methods *Questions of theory and practice of geological interpretation of geophysical fields: Materials of the 39th session of the International Scientific Seminar named after D. G. Uspensky* (Voronezh) pp.162-165
[13] Borovsky M Ya *et al 2015 Geophysical prospecting for the detection of hydrocarbons in shale strata on the territory of the Republic of Tatarstan* (*Oil industry* **11**) pp 94-96.
[14] Borovsky M Ya 2015 Non-seismic methods for studying dominant deposits *Prospects for increasing the resource base of the developed deposits, including those from the house deposits.* (Almetyevsk: PJSC Tatneft) pp 191-199.
[15] Uspensky B V *et al 2016 Formation conditions and geophysical methods for forecasting complicating factors for the development of deposits of natural bitumen* (*Oil Industry* **11**) pp 75-77
[16] Khisamov R S *et al 2011 Geological and geophysical prospecting of oil deposits at a late stage of development* (Kazan: FEN Publishing House of the Academy of Sciences of the Republic of Tatarstan) p 228

Institute of Physics
Dirac House, Temple Back
Bristol BS1 6BE UK

ISSN: 1755-1307
ISBN 978-1-5108-6338-5

9 781510 863385

CURRAN ASSOCIATES INC.
proceedings.com

9th International Conference on Environmental Science and Development (ICESD 2018)

IOP Conference Series: Earth and Environmental Science
Volume 151

Paris, France
7 – 9 February 2018

IOP
Institute of Physics